Climate Change
and
Food Production Challenges

Climate Change
and
Food Production Challenges

Vikram Sharma

RANDOM PUBLICATIONS

NEW DELHI (INDIA)

Climate Change and Food Production Challenges

ISBN 978-93-5111-503-8

Published in 2015 in India by

RANDOM PUBLICATIONS
4376-A/4B, Gali Murari Lal, Ansari Road
New Delhi-110 002
Phone : +9111-43580356, 011-23289044, 011-43142548
e-mail: sales@randompublications.com,
info@randompublications.com, randomexports@gmail.com

Reprinted 2022

Type Setting by : Friends Media, Delhi-110089
Digitally Printed at : Replika Press Pvt. Ltd.

Preface

Climate change and food production are interrelated processes, both of which take place on a global scale. Global warming is projected to have significant impacts on conditions affecting agriculture, including temperature, carbon dioxide, glacial run-off, precipitation and the interaction of these elements. These conditions determine the carrying capacity of the biosphere to produce enough food for the human population and domesticated animals. The overall effect of climate change on food production will depend on the balance of these effects. Assessment of the effects of global climate changes on agriculture might help to properly anticipate and adapt farming to maximize agricultural production.

Despite technological advances, such as improved varieties, genetically modified organisms, and irrigation systems, weather is still a key factor in agricultural productivity, as well as soil properties and natural communities. The effect of climate on agriculture is related to variabilities in local climates rather than in global climate patterns. The Earth's average surface temperature has increased by 1.5°F since 1880. Consequently, agronomists consider any assessment has to be individually consider each local area. On the other hand, agricultural trade has grown in recent years, and now provides significant amounts of food, on a national level to major importing countries, as well as comfortable income to exporting ones. The international aspect of trade and security in terms of food implies the need to also consider the effects of climate change on a global scale.

This book analyses the direct impacts of climate change on food production, and the challenges faced by agricultural sector. This thorough and timely volume is an invaluable resource for anyone interested in exploring the impacts of climate change in arguably the most important sector of the world economy.

Author

Contents

1

Relevance of Climate to Food Production

Climate fluctuates naturally on all time scales, from days and years (variability proper) to decades and centuries (climate change). FAO has considerable experience in dealing with the first, which is a major cause of the variability of inter-annual agricultural production. It has less experience, however, in the area of climate change and needs strengthening to handle the many new challenges and opportunities arising under international climate change negotiations.

Since agriculture is now seen as a part of the solution to the climate change problem, there is even the possibility that new resources, for the agriculture sector at the national level and for sustainable rural development in general, will become available through the Kyoto Mechanisms during climate negotiations. Since climate affects almost all sectors of agriculture, a proper and coordinated climate programme should be developed to support FAO activities. This should focus on policy formulation, data, capacity building, and collaboration with the Secretariat of the United Nations Framework Convention on Climate Change (UNFCCC) and the Intergovernmental Panel on Climate Change (IPCC).

It should also strengthen several fields of renewed relevance in the climate change context, such as sequestering carbon, reducing green-house gas emissions, substituting fossil fuels by biofuels, and generally improving the resilience of agriculture to climate variability and change.

Climate Negotiations And Involvement of FAO

Recognising the problem of potential global climate change, the World Meteorological Organization (WMO) and the United Nations Environment Programme (UNEP) established the Intergovernmental Panel on Climate Change (IPCC) in 1988. During the same year, FAO created an *ad hoc* Interdepartmental Working Group on Climate Change and Variability in Relation to Food Security.

IPCC stated, in 1995, that there was a discernible effect of man on climate1, through the emission of greenhouse gases, mainly carbon dioxide (CO_2), methane (CH_4) and nitrous oxides (NOx), which together are responsible for most of the warming effect. It is generally agreed that about one quarter of the main greenhouse gas (CO2) stems from agricultural sources (land-use change, deforestation and biomass burning). CH4 accounts for a smaller fraction of the warming effect but most of it comes from agricultural sources such as domestic ruminants, forest fires, wetland rice cultivation and waste products. Conventional tillage and conventional fertiliser use together are the source of 70% of the NOx. Altogether, agricultural sources are responsible for about 30% of global warming.

In 1996, several UN Agencies, including FAO, established the Inter-Agency Committee on the Climate Agenda (IACCA) to coordinate their climate activities in the wake of Agenda 21. The 111th session of the FAO Council unanimously supported the association of FAO with this initiative.

The World Food Summit (WFS), organized by FAO in 1996, and in particular the WFS Plan of Action (WFSPA) underlined that the resource base for food, agriculture, fisheries and forestry was under stress and threatened by problems such as desertification, deforestation, overfishing, loss of biodiversity, inefficient use of water and climate change. Climate change itself can exacerbate climate variability, which remains one of the main factors behind the inter-annual instability of food production. Climate variability includes altered frequencies of extreme events.

With the aim of countering the environmental threats to food security, the WFSPA encouraged governments to take the anticipated impacts of natural climate variability and climate change on rainfall and temperature into account in developing agricultural and land-use policies.

The WFSPA further called for the early implementation of the 1992 UNFCCC, which entered into force in March 1994. The main objective of

UNFCCC is to stabilise greenhouse gas concentrations in the atmosphere at a level that would prevent dangerous anthropogenic interference with the climate system.

The UNFCCC was supplemented in 1997 by the Kyoto Protocol (KP), which is generally expected to enter into force around 2002. The agreements commit most Annex I countries2 to quantified reductions of their greenhouse gas emissions below 1990 values during the first commitment period (2008-2012). Developing countries are not currently bound by any reduction targets, but the KP makes provision for several mechanisms (the Kyoto Mechanisms) that will foster sustainable development through collaboration between developed and developing countries.

Effects of Climate on Food Security

There are many interactions between climate variability and change, and agriculture. Agriculture is affected by the vagaries of climate, and contributes to increasing climate variability and change, directly and indirectly, through the emission of greenhouse gases, the disruption of natural cycles of many elements and water through land degradation and deforestation, etc. In addition to considering the technical and scientific aspects, the international community is now taking measures, mostly in the policy, legal and institutional spheres, which will commit countries to adopting new practices and legislation affecting the agriculture and forestry sectors.

Because the patterns of climate variability are modified only gradually, climate change can go un-noticed for some time. Climate variability *per se* is not necessarily harmful; the problems arise from extreme events, and the uncertainty which derives from the difficulty of predicting weather beyond a week or so.

The effects of climate variability on all forms of agricultural production are well known. Depending on the level of development and technological influences on yields, 10% to 100% of the short-term variability of production can be ascribed to vagaries of the weather. The losses associated with the background variability of climate are significantly higher than those associated with spectacular but localised weather-related disasters like cyclones and flooding.

The possible effects of climate change are still being debated in the scientific community. Ignoring hypothetical impacts, the followingglobal changes are described as "very probable to virtually certain3" between now and 2100 (local values can be significantly larger or smaller than the average figures):

1. average temperature increase of about 2° C. This corresponds to a latitudinal shift of climates of about 200 km towards the poles, with a reduction of permafrost areas;
2. global mean precipitation increase of up to 5%, and intensification of the water cycle (a faster turnover rate of water in the atmosphere: increased evapotranspiration rates due to higher temperatures, and concentration of rainfall into a smaller number of rainy events);
3. average sea level rise of about 50 cm;
4. whatever changes will take place, they will persist for decades or centuries, as climate has considerable inertia.

Potential impacts of climate change on agricultural production will depend not only on climate *per se*, but also on the internal dynamics of agricultural systems, including their ability to adapt to the changes. In the words of Reilly (1996), *the most "robust conclusion" that does emerge from the studies is that climate change has the potential to change the productivity significantly.* In comparison with global impacts, impacts at the level of plants and animals are relatively easy to assess:

1. a lengthening of farming and freshwater fishing seasons;
2. shorter cycles for all organisms, including pests;
3. increased respiration loss due to higher night-time temperatures;
4. CO_2 fertilisation, with a more marked effect on maize, sorghum, millet and sugarcane;
5. improved water-use efficiency of crops;
6. modifications of coastal/delta agriculture;
7. modified crop/animal and pest/disease relationships, including new pests and diseases, and changes in economic return (including new opportunities);
8. modified variability and risk patterns.

Global impacts of climate change on food production and food security may include:

1. marked changes in the geographic distribution of climates and their associated land-use patterns, possibly resulting in a modification of the

crop geopolitical balance, including a potential net positive impact on production in developed temperate countries and a negative impact in tropical developing countries;

2. loss of carbon stored in peat and soil organic matter;
3. modified erosion patterns and increased land degradation;
4. human population movements and increased global insecurity;
5. loss of biodiversity, and modification of ecosystem species composition and metabolism;
6. zonal migration of species, ecosystems, crops and animals;
7. modification of the interactions and balance between species, including pests and diseases.

FAO Activities Related to Climate

Formal climate-related work started in 1968 when FAO, WMO and UNESCO established the Interagency Group on Agricultural Biometeorology, which was later joined by UNEP. After the 1972 Stockholm Conference on the Human Environment and the 1974 World Food Conference, FAO launched two activities with a major climate component: the Agro-Ecological Zones Project (coordinated by the Land and Water Development Division) to estimate the food production potential of developing countries, and the Global Information and Early Warning System on Food and Agriculture to continuously monitor the crop prospects and food situation throughout the world and alert the international community of impending serious food shortages. Both have led to the development of climate databases and methods for assessing the impact of climate on agriculture (Environment and Natural Resources Service, SDRN) for planning (Land and Water Development Division, AGL) and monitoring purposes (Commodities and Trade Division, ESC).

FAO's *ad hoc* Working Group on Climate, established in 1988 is currently chaired by the Director of the Forest Products Division (FOP), while SDRN handles the secretariat. All technical Departments and the Legal Office participate in the Group.

A number of documents directly or indirectly related to climate and climate change have been produced, in particular *FAO position papers on climate change* which were distributed to major meetings such as the UN General Assembly, the Second World Climate Conference, UNCED, the

World Food Summit and various international gatherings like the Conference of the Parties to UNFCCC.

FAO has participated regularly in the activities of the WMO Commission for Agricultural Meteorology (CAgM). The Organization has also organized and participated in a number of international fora and conferences with climate change and sustainable development as core issues, such as IPCC meetings, the UNFCCC Subsidiary Body for Science and Technological Advice (UNFCCC/SBSTA), and in particular the recent (June 2000) sessions that discussed the Special Reports on Land Use, Land Use Change and Forestry (SRLULUCF).

In addition to these activities, which are largely meant to serve government planners, FAO's work is also directed at:

(1) improving knowledge of issues related to climate change (the 1990 Expert Consultation on Forestry and Climate Change, the 1993 Expert meeting on Global Climate Change and Agricultural production, and the 1993 Expert Consultation on Biofuels for Sustainable Development), and

(2) improving farmers' capacities to reduce risk or make optimal use of climate resources (rainfall, sunshine, etc.) by better understanding their variability.

More recently, the Organization has also examined the complex relationships between land degradation, carbon sequestration and biodiversity in a joint FAO/IFAD Expert Meeting (1999), prepared documents on *El Protocolo de Kyoto y el Mecanismo para un Desarollo Limpio - Nuevas Posibilidades para el Sector Forestal de América Latina y el Caribe* - (1999), *Climate Change and the Forestry Sector: Possible Legal Responses and Their Implications* (2000), and assessed the relevance of the Verification of Country Level Carbon Stocks and Exchanges for developing countries in an expert meeting (September 2000). The Organization co-sponsored the Expert Workshop on*Carbon Sequestration, Sustainable Agriculture and Poverty Alleviation* in WMO (Geneva) in August-September 2000.

UUFCCC and Kyoto Protocol Provisions

While the UNFCCC makes only brief mention of agriculture and forestry, the Kyoto Protocol deals explicitly with the subject and puts special emphasis on sustainable management practices and the *promotion of sustainable forms of agriculture in the light of climate change considerations.* Land-use changes, enteric fermentation, manure management, rice cultivation, agricultural soils, prescribed burning of savannahs and field burning of agricultural residues are all listed as sources of greenhouse gases which will have to be taken into consideration by countries in their reports on anthropogenic sources and sinks to the Conference of the UNFCCC Parties.

Several articles of the KP are of particular relevance to the agricultural community and Members of FAO. They include the sections on Afforestation, Deforestation and Reforestation (Article 3.3) as well as other sources and sinks, including soil and biomass storage, (Article 3.4) and their estimation (Article 5). Also particularly relevant are tradable permits5 (Article 6), annual reporting (Article 7), review and control mechanisms (Article 8), improved methodologies to assess emissions and sinks (Article 10), and especially the Clean Development Mechanism (CDM) which is described in Article 12.

Article 10 commits countries to formulating national and regional programmes to improve scientific cooperation, education and training programmes, inventories of emissions and sinks, and to improve mitigation as well as adaptation potential. Article 11 makes provision for developed Parties to assist developing ones. This includes, in particular, reporting commitments and technology transfer (Article 4 of the UNFCCC).

The double purpose of the CDM is to assist developing Parties in achieving sustainable development and developed Parties in achieving compliance with their emission limitations and reduction commitments. The CDM will allow Annex I Parties and Non-annex I Parties to jointly implement projects which will result in certified emission limitations. This mechanism complements the tradable permits approach between Annex I Parties. While details of the implementation of the CDM are still being debated, in particular as regards the fossil fuel substitution by biofuels and the admissibility of soil carbon pools as sinks, its potential as a sustainable development mechanism has to be carefully assessed.

The active participation of international organizations in international climate issues is clearly stressed in the KP. UNFCCC lists several categories of country which will require specific assistance, including: small island countries, countries with low-lying coastal areas, countries with arid and semi-arid areas, forested areas and areas liable to forest decay, countries with areas prone to natural disasters, countries with areas liable to drought and desertification, and eventually countries with areas with fragile ecosystems, including mountainous ecosystems.

Role of FAO in Climate Change Negotiations

FAO and its members are faced with several challenges directly or indirectly deriving from the current climate negotiations. These include

(1) new obligations (commitments),

(2) new opportunities linked with the fact that carbon has now become a new 'commodity' which has to be monitored, quantified and sometimes managed differently than in the past,

(3) a renewed raison d'être for several activities in which the Organization has been involved and which are being requested by the climate change community, and

(4) some areas which need strengthening.

Under the KP, Parties have to formulate, implement, publish and regularly update national and, where appropriate, regional programmes containing measures to mitigate climate change and facilitate adaptation to climate change in the energy, transport and industry sectors as well as agriculture and forestry.

New Obligations

Inventories of sources and sinks: the UNFCCC commits all signatories to carrying out detailed inventories of anthropogenic sources of greenhouse gases (strictly those greenhouse gases not covered by the Montreal Protocol). The KP is more specific in that it states that countries are committed to verifiable changes in carbon stocks, including those deriving from land-use changes and which are of direct relevance to the agricultural community. FAO statistics in general, while not detailed enough for many of the new reporting commitments, such as carbon distribution in sequestration projects or GHG emissions by type of activity, still provide a useful benchmark and a historical perspective which other data lack. FAO statistics have to be re-examined, in particular in the light of the new reporting commitments and national carbon accounting.

Mitigation of emissions of agricultural greenhouse gases is largely compatible with a no-regrets approach (often termed win-win situations) since most emissions from the agricultural sector are indicators of inefficient and sometimes unsustainable farming. Such emissions constitute losses to farmers and there is room for significant reduction in many instances.

New Opportunities

Many activities have been carried out by the farming and forestry communities to increase the sustainability of production systems. These, listed below, may now qualify for one of the Kyoto Mechanisms, and as such earn carbon credits, thereby increasing farmer income or attracting investment benefiting sustainable development and the population at large. Due attention should also be paid to the ecological/ecosystem functions of carbon when assessing the scope and relevance of 'carbon' activities in agriculture.

The activities include:

1. improved fertiliser use, since N_2O released into the atmosphere is a loss and an indication of inefficient farming;
2. improved ruminant digestion through more efficient feeds or, when feasible, a shift to enzymic digesters;
3. development of water harvesting and conservation techniques, and other improvements to crop-water management, as an adaptation to rainfall variability;
4. improved rice farming, since higher yields are accompanied by a relatively smaller loss of methane;
5. conservation agriculture and practices leading to improved soil carbon storage (carbon sink) while at the same time improving soil structure and increasing water holding capacity;
6. improved low-impact harvesting in forests, reduction of slash-and-burn agriculture, better soil protection.

Renewed Raison D'être

The Global Terrestrial Observing System (GTOS) was established jointly by FAO, ICSU, UNEP, UNESCO and WMO in 1996 to provide policy makers, resource managers and the research community with the long-term data they need to detect global change and to assess the capacity of terrestrial ecosystems to support sustainable development. The secretariat of GTOS is located in FAO. One of its priority areas is climate change and GTOS is now leading the development of a Terrestrial Carbon Observation Initiative (TCOI) which is aimed at assisting countries to assess and monitor the carbon sources and sinks in the agricultural and rural sectors of their countries through the systematic collection of carbon-related data using ground and satellite observations. TCOI is intended to fill data and information gaps in the terrestrial carbon cycle and build globally-consistent carbon data sets that will eventually be required under the UNFCCC. Both activities fill a gap in the data requirements of the UNFCCC.

A broad range of data is collected and maintained by FAO, which is of direct relevance not only to the climate change problem in general, but also to the more immediate operational (reporting) requirements of countries under the UNFCCC. Unlike the above-mentioned national and sub-national agricultural statistics, these data are mostly geo-referenced. They include:

1. all information on vegetation, soil and soil organic matter (source and sink of carbon dioxide);
2. climate data and maps;
3. biomass indicators derived from satellite remote sensing;
4. fertiliser use (source of greenhouse gases);
5. areas cultivated under rice and ruminant cattle statistics (source of methane); and
6. use of inputs, like methyl bromide.

Bio-energy and fossil fuel substitution by biofuels. The substitution of carbon from fossil fuels by biofuels and other renewable forms of energy are a most sustainable approach to reducing carbon dioxide emissions. Global climate change and the realisation of the potential of biomass energy as a substitute for fossil fuels have triggered new attention on the energy function of agriculture. Currently biomass is the main source of energy in many developing countries and provides about 15% of the world's energy, but this could be increased significantly through modern conversion technology. Much of this energy comes from various types of agricultural and forestry residues, although in the future various types of energy crops and plantations are expected to provide the main source. Since the early 1990s, the increasing interest in biomass for energy has been manifested in most energy scenarios showing biomass as a potential major source in the 21st century. The expected increase could have a significant impact for mitigating CO_2 emissions, sequestering carbon and increasing employment for many farmers in remote rural areas.

Areas which Need Strengthening

FAO does not currently have a clear climate change programme, although the AG, ES, FO, TC and SD Departments have been working on climate variability issues for decades. New emphasis on climate change has been covered on an *ad hoc* basis but suffers from the lack of a coordinated programme that would encompass climate variability and change, and allow the Organization be more proactive at all levels, from the international scene to the individual farm.

Capacity building stands high on the agenda of all international climate meetings and negotiation fora, starting with the Conference of the Parties (CoP) to the UNFCCC. FAO does not currently have the capacity to respond to the requests coming from the international and national climate community. FAO has experience in preparing training programmes, and developing and disseminating appropriate techniques which can be used to improve the capacity for dealing with climate change issues of decision makers, national experts responsible for reporting on carbon issues, and agricultural personnel at all levels. FAO could also take a more active role in ensuring that climate variability and change issues are well covered in the curricula of agricultural training institutions at all levels.

More research will be needed in the development of measures for adaptation to climate change, the application of existing knowledge and the systematic assessment of anticipated impacts of climate change in agriculture. This includes prevention, protection, impact assessment and rehabilitation in the case of extreme climatic events, as well as drought and desertification impact reduction, especially in Africa. Much of the work of FAO derives directly from the need to adapt farmers' practices to adverse conditions and to help stabilise and possibly increase production under a very variable environment. As such it responds directly to UNFCCC concerns. An important option is Response Farming (a set of techniques to optimise on-farm decision making by closely monitoring crop growing conditions, in particular weather, and modelling the combined impact of environment and management), modernised through the use of current satellite and ground observation and data transmission techniques. Other immediate options include improved use of weather and climate forecasts; early detection of epizootic diseases based on satellite images and climate prediction, better crop forecasts, and better farming practices.

Policy formulation: FAO should strengthen its capacity to assist countries to take climate change considerations into account in the formulation of their national policies, not only in the field of environment but also in the broader development of socio-economic policy. This includes the promotion of policies that create incentives for land users to limit emissions or generate sinks, as well as the development of bioenergy and the adjustment of the institutional and legal context to improve the resilience of production systems to climate variability and change. Where necessary, FAO should be in a position to assist in the formulation and harmonisation of regional policies.

FAO should develop closer and more systematic collaboration with the UNFCCC secretariat and with IPCC regarding a number of issues, starting with the crucial issue of definitions, which are extremely important in the international legal framework. The Organization has to ensure that the requirements of the agricultural community are duly taken into account in the international negotiations, and that the adopted measures and policies are realistic, feasible and equitable. FAO should also encourage synergies and collaboration in the agricultural domain between the UNFCCC and IPCC and the other environmental conventions, in particular those on desertification (UNCCD) and biodiversity (CBD).

FAO should take an active role in several aspects of the collection, verification, standardisation and dissemination of the data required for the implementation of the UNFCCC. This specifically includes the formulation of agricultural statistical methods compatible with the reporting requirements of the UNFCCC, the Forest Resources Assessment, soil carbon monitoring, GHG emission inventories, the Global Terrestrial Observing System (GTOS) and particularly the Terrestrial Carbon Observation Initiative (TCOI).

COAG Guidance

COAG guidance is requested on how to promote and take advantage of the fact that agriculture constitutes one of the solutions to current concerns about climate change. This includes both technical and institutional aspects, which are of relevance to most Departments of FAO. In particular, guidance is sought on the following issues and proposals that have direct implications for FAO's current and future activities in this field:

1. Development of an integrated FAO climate-change programme based on current activities as well as the new opportunities arising under the international climate negotiations. This should include:
2. the promotion of agriculture as a key player in the reduction of atmospheric greenhouse gases and the resulting threats associated with climate change, including the active participation of the Organization in international climate discussions, as well as capacity building and raising the awareness of agricultural personnel and decision makers;
3. the identification, development and promotion of practices which will reduce emissions of greenhouse gases or sequester carbon, while contributing to sustainable development and assisting members in identifying opportunities for earning carbon credits under the Kyoto mechanisms;
4. the identification and strengthening of FAO's activities with a potential for assisting Members with (1) their reporting, (2) mitigating emissions of greenhouse gases, (3) increasing the resilience of production systems against the vagaries of current climate and the threats of climate change, (4) improving their adaptation capacity. This includes conservation agriculture, biomass and soil storage of carbon, improved use of fertiliser and improved ruminant digestion through improved feeds, as well as non-structural measures such as crop insurance linked to more sustainable practices and the promotion of bioenergy as a sustainable substitute for fossil fuels;
5. active participation and, where appropriate, a leading role in the collection, adaptation, exchange and dissemination of geo-referenced and tabular climate-change data and statistics for agriculture. This includes the development of information, databases, and networks to monitor soil carbon, strengthening the role of the Global Terrestrial Observing System and the Terrestrial Carbon Initiative, and the collection of data required for carbon accounting at the national and sub-national level;
6. improved use of tools to assess the impact of atmospheric conditions on agricultural production (crop forecasting), the use of weather and climate forecasts in farm-level decision-making, and the development

of techniques which optimise the use of climate resources under variable and sometimes extreme conditions.

7. Stronger capacity in the Organization to assist members (1) to assess and monitor climate variability and change impact on agriculture, and (2) to adopt appropriate national policies and international negotiating positions.
8. Improvement of the inter-departmental coordination mechanism through a formal inter-departmental Group, with terms of reference reflecting the points above, as well as proper resources.

References

Bazzaz, F., and W. Sombroek (eds.), *Global climate change and agricultural production. Direct and indirect effects of changing hydrological, pedological and plant physiological processes.* FAO and John Wiley & Sons, 345 pp.

Houghton, J.T., L.G. Meira-Filho, B.A. Callander, N. Harris, A. Kattenberg and K. Maskell (Editors), 1995. *Climate Change 1995, the science of climate change.* Cambridge University Press, 572 pp.

Mahlman, J.D., 1997. Uncertainties in projections of human-caused climate warming. *Science,* 278:1416-1417.

2

Climate Change-Agriculture Conundrum

The risks associated with climate change lie in the interaction of several systems with many variables that must be collectively considered. Agriculture can be defined as one of the systems, and climate the other. If these systems are treated independently, this would lead to an approach which is too fragmentary. The issue is more global. It is now held as likely that human activities can affect climate, one of the components of the environment. Climate in turn affects agriculture, the source of all food consumed by human beings and domestic animals. It must be further considered that not only climate may be changing, but that human societies and agriculture develop trends and constraints of their own which climate change impact studies must take into consideration.

World Agricultural Context

In general, global food production has been growing faster than human population (Table 1). There are, however, marked disparities between continents. For example, in Africa local production of cereals cannot keep pace with population increase, and production of root and tuber crops is growing faster than that of the nutritionally more valuable cereals. In contrast, arable land growth lags behind population growth, which indicates some intensification of production. In Asia, the upper limit of available land has been reached in several countries, resulting in very high cropping intensities and a dominant role for irrigation. In Latin America, the increase in arable land is achieved only at a high ecological cost (especially deforestation) which may have direct relevance to climate change.

Table 1. Some trends in worldwide population, food production, arable and irrigated land. Europe and Asia do not include the former USSR. Arable land is dermed as all agricultural land, excluding perennial crops and permanent rangeland. Irrigated land is in % of the land under annual and permanent crops, average value from 1981 to 1990. Fertilizer use is the 1980-1990 average

Continent	1961-1990 exponential growth rate (%)					% Cereal yield increase (1961-1970 and 1986-1990)	Fertilizer use (kg fertilizer per ha arable land)	Irrigated land
	Population	Production			Arable land			
		Cereals	Tubers	Pulses				
Africa	2.83	1.95	2.74	1.58	0.44	32	1.2	5.8
North and Central America	1.52	2.32	1.30	1.53	-0.59	49	10.7	9.6
South America	2.34	2.80	0.51	0.71	1.53	44	2.6	5.8
Asia	2.11	3.30	1.95	0.26	0.07	86	14.5	31,0
Europe	0.51	2.37	-1.27	1.57	-0.53	75	64.3	10.8
Oceania	1.72	3.14	1.62	13.76	1.23	34	2.0	4.1
World	**1.88**	**2.66**	**0.78**	**0.83**	**0.12**	**65**	**9.6**	**15.2**

According to a recent FAO prospective study covering the years 1988/1990 to 2010, 'the rate of growth of agricultural land will be further reduced during the next two decades. The pressures on *fresh* water *resources,*however, will be considerable, as will be those on the environment arising from the intensification of land use'.

A significant reduction in world population growth rates is foreseen from 1.8% (1980/1990) to 1.4% (2000/2010), roughly equivalent to an increase of the population doubling time from 40 to 50 years. The projected continuation of the high population growth in Africa can be related to slow economic development during the coming two decades, since in general a reduction of poverty precedes reduction of population growth.

AT-2010 also lists the following significant trends for the near future:

(i) world production of cereals will continue to grow, but not in per capita terms;

(ii) export cereals will undergo a modest growth in demand;

(iii) the livestock sector in developing countries will continue to grow;

(iv) root crops, tubers and plantains will retain their importance;

(v) oil crops will undergo rapid growth in developing countries; and, most importantly,

(vi) many developing countries will become net agricultural importers.

As indicated above for population increase, large differences exist in the rate of agricultural development among the developing continents, while differences among the developed countries tend to level out.

It is also likely that hunger and under-nutrition, which currently affects 800 million people (about 20% of the 4 000 million inhabitants of developing countries), will be reduced, but large pockets of malnutrition will persist. Pressure on environmental resources will continue to build up.

Changing Agricultural Environment

Climate constitutes a complex of inter-related variables. On average, through a set of regulatory mechanisms, a smooth change in one variable triggers smooth changes in most others. With the exception of possible qualitative and abrupt variations, which will be mentioned below, such inter-relations are independent of atmospheric carbon dioxide (CO_2). The latter and other greenhouse gases play a part largely through their effect on the radiation balance of the atmosphere.

There is only a weak link between such factors as cloudiness and wind. Temperature, evaporation and rain are strongly correlated, which illustrates the likely intensification of the hydrological cycle (Figure 1). Combined with the projected pressure on land and water use, competition for land and water will certainly become a key social and political issue.

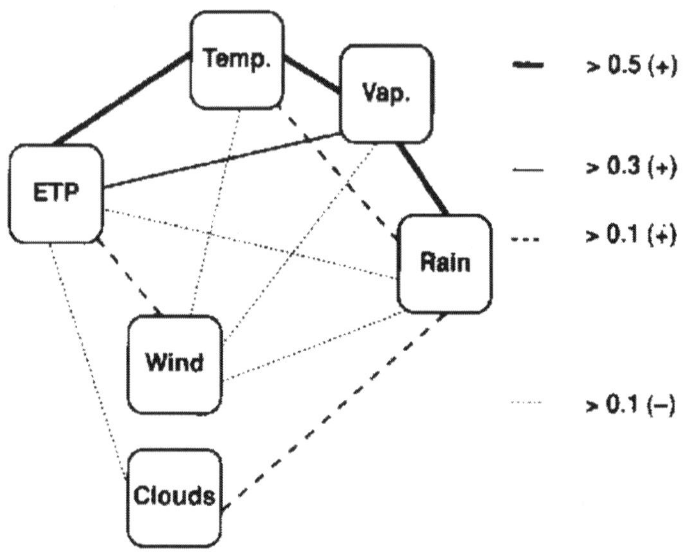

Figure 1. Some relationships between major climate variables (average temperature, Temp.; water vapour pressure, Vap.;rainfall, Rain; wind speed, Wind; cloudiness, Clouds; and evapo-transpiration potential, ETP). Solid and dotted lines indicate positive and negative correlations, respectively. The strength of the correlation decreases from double heavy lines to thin single lines.

Climate variability is likely to increase under global warming, both in absolute and in relative terms. This is linked with thresholds which affect the occurrence of many meteorological phenomena. For instance, tropical cyclones are 'fed' by water vapour evaporating from oceans at a temperature above 26 or 27°C. Therefore, higher average sea surface temperatures are bound to result in a higher frequency of tropical cyclones.

The rate of change itself is extremely important. For example, recent work on the saw-tooth temperature changes in the past as observed in the Arctic, raise concern that changes may occur abruptly, with average temperatures changing by 10 or 12°C in just a matter of decades. The mechanism of such changes is not clear as yet, but seems to involve the mechanical stability of ice sheets, and sudden changes in the hydrological cycle and the Atlantic conveyor current.

Such changes would, of course, be associated with dramatic changes in the distribution and quantities of ocean products, and cause havoc to established national fishery activities. They would also make adaptation to climate change, together with most agricultural planning, extremely difficult.

Agricultural Greenhouse Gases

In addition to water vapour, important greenhouse gases are carbon dioxide (CO_2), methane (CH_4), nitrous oxide (N_2O), tropospheric ozone (O_3) and chlorofluorocarbons (CFCs). The basic characteristics of the first three gases are given in Table 2. The degree to which these greenhouse gases stem from agricultural sources is also given. The exact ratio between these land-use related emissions and those from natural ecosystems (swamps, tundra), fossil fuels (coal, oil, gas) and geological sources (volcanoes) are a bone of contention among industrialized, oil producing and developing countries.

Moral rights and duties in relation to the international flow of development and conservation funds are involved. Therefore, better estimates based on exact measurements of the net greenhouse gas emissions from agricultural practices in developing countries are urgently required to assess responsibilities properly. The required reductions of emissions to achieve stabilization of atmospheric concentrations of current levels are believed to be >60% for CO_2 15-20% for CH_4 and 70-80% for N_2O. This implies the need for up-to-date and complete information on land cover and land uses per country: not only the kind of crops grown but also their intensity, their rotation and the amount of inputs (energy, fertilizers).

The current trends of some of the agricultural sectors directly associated with greenhouse gas emission sources are listed in Table 3. The loss of 'natural land', including tropical forests to agriculture, grazing, logging and urbanization, may continue, though at a slower pace (*AT-2010*). Many earlier estimates are now regarded with suspicion, as the borderline between crop agriculture, forest and cattle agriculture appears to be a fuzzy one.

Finally, both recent trends of fertilizer use and *AT-2010* projections indicate major increases in consumption. Even with appropriate measures to optimize fertilizer use, it is likely that N_2O losses from fertilizers will continue to increase.

Table 2. An overview of agricultural greenhouse gases with the trends as currently envisaged. ppmv and ppbv stand for parts per 10^6 and parts per 10^9, respectively, by volume.

	CO_2	CH_4	N_2O
Atmospheric lifetime (yr)	120	14.5	120
Direct GWP [1]	1	24.5 [2]	320
Pre-industrial concentration [3]	280 ppmv	0.8 ppmv	288 ppbv
Present-day levels	360 ppmv	1.72 ppmv	310 ppbv
Current annual increase (%)	0.5	0.9	0.25
Major agricultural sources [4]	deforestation	- wetland rice	- synthetic N fertilizers
		- ruminants	- animal excreta
		- biomass burning	- biological N fixation
Percentage of global source stemming from agriculture	30	40	25
Predicted change 1990-2020	-	+	+

[1] GWP, Global Warming Potential, is the direct warming effect in relation to CO_2 at a time horizon of 100 years.

[2] Includes indirect effects through chemistry.

[3] About year 1750-1800.

[4] Activities responsible for emissions are projected to increase by: rice (+ 10%); ruminant population (+30%); synthetic fertilizer use (+20%): animal excreta and biological N fixation increasing but rate not specified.

Table 3. Growth rates between 1961 and 1990 in agricultural sectors responsible for greenhouse gas emissions. Europe and Asia do not include the former USSR. Domestic ruminant numbers were computed as the sum of cattle, sheep, goats, camels and buffaloes

Continent	1961-1990 exponential growth rate (%)			
	Ruminant numbers	Forested area	Rice area	Fertilizer consumption
Africa	1.29	-0.43	2.23	6.21
N and C America	-0.07	-0.02	1.50	3.29
S. America	1.29	-0.49	1.65	9.15
Asia	1.18	-0.59	0.62	9.54
Europe	0.33	0.25	0.78	2.75
Oceania	0.07	-1.15	5.88	1.25
World	**0.90**	**-0.26**	**0.74**	**5.35**

Ecological and Indirect Climate Effects

In qualitative terms, many indirect effects of climate change on agriculture can be conjectured. Most of them are estimated to be negative and they catch most of the attention of the media. These effects include:

1. the overall predictability of weather and climate would decrease, making the day-to-day and medium-term planning of farm operations more difficult;
2. loss of biodiversity from some of the most fragile environments, such as tropical forests and mangroves;
3. sea-level rise (40 cm in the coming 100 years) would submerge some valuable coastal agricultural land;
4. the incidence of diseases and pests, especially alien ones, could increase;
5. present (agro) ecological zones could shift in some cases over hundreds of kilometres horizontally, and hundreds of metres altitudinally, with the hazard that some plants, especially trees, and animal species cannot follow in time, and that farming systems cannot adjust themselves in time;

6. higher temperatures would allow seasonally longer plant growth and crop growing in cool and mountainous areas, allowing in some cases increased cropping and production. In contrast, in already warm areas climate change can cause reduced productivity;
7. the current imbalance of food production between cool and temperate regions and tropical and subtropical regions could worsen.

Plant Physiological Direct Effects

The greenhouse gases CH_4, N_2O and chlorofluorocarbons (CFCs) have no known direct effects on plant physiological processes. They only change global temperature and are therefore not discussed further. Instead, concentration should be on the effects of increased CO_2tropospheric O_3, increased UV-B through depleted stratospheric ozone, increased temperatures and the associated intensification of the hydrological cycle.

The CO_2 Fertilization Effect

CO_2 is an essential plant 'nutrient', in addition to light, suitable temperature, water and chemical elements such as N, P and K, and it is currently in short supply.

Higher concentrations of atmospheric CO_2 due to increased use of fossil fuels, deforestation and biomass burning, can have a positive influence on photosynthesis; under optimal growing conditions of light, temperature, nutrient and moisture supply, biomass production can increase, especially of plants with C3 photo-synthetic metabolism, above and even more below ground

A total of 10 to 20% of the approximate doubling of crop productivity over the past 100 years could be due to this effect and forest growth or regrowth may have been stimulated as well. Further productivity increases may occur in the coming century, in the order of 30% or more where plant nutrients and moisture are adequate.Higher CO_2 values would also mitigate the plant growth damage caused by pollutants such as NO_x and SO_2 because of smaller stomatal openings. Higher percentages of starch in grasses improves their feeding quality, implying less need for feed mixes when silaging.

The CO_2 Anti-transpirant Effect

With increased atmospheric CO_2 the consumptive use of water becomes more efficient because of reduced transpiration. This is induced by a contraction of plant stomata and/or a decrease in the number of stomata per unit leaf area. This restricts the escape of water vapour from the leaf more than it restricts photosynthesis. With the same amount of available water, there could be more leaf area and biomass production by crops and natural vegetation. Plants could survive in areas hitherto too dry for their growth.

Ultraviolet Radiation

Increased ultraviolet radiation (UV-B, between 280 and 320 nanometres), due to depletion of the stratospheric ozone layer, mainly in the Antarctic region, may negatively affect terrestrial and aquatic photosynthesis and animal health. Over the last decade, a decrease of stratospheric ozone was observed at all latitudes (about 10% in winter, 0% during summer and intermediate values during spring and autumn). However, the 'Biological Action Factor' of UV-B can vary over several orders of magnitude with even slight changes in the amount and wavelength of UV-B.

Tropospheric Ozone

Tropospheric ozone originates about half from photochemical reactions involving nitrogen oxides (NO_x), methane or carbon monoxide, and half by downward movement of stratospheric ozone.

High ozone concentrations have toxic effects on both plant and animal life. It is likely that ozone, in conjunction with other photo-oxidants, is contributing towards the 'new type of forest damage' observed in Europe and the United States.

In the tropics, tropospheric ozone concentrations are generally lower than at northern mid-latitudes. However, this does not apply to periods when biomass burning releases precursor substances for the photochemical formation of ozone.

Temperature

In general, higher temperatures are associated with higher radiation and higher water use. It is relatively difficult to separate the physiological effects (at the level of plants and plant organs) of temperatures from the ecological ones (at the level of the field or of the region). There are both positive and negative impacts at the two levels, and only crop- and site-specific simulation can assess the global 'net' effect of temperature increases. It is generally agreed that:

1. rising temperatures - now estimated to be 0.2°C per decade, or 1 °C by 2040 with smallest increases in the tropics (IPCC, 1992) - would diminish the yields of some crops, especially if night temperatures are increased (the temperature increase since the mid-1940s is mainly due to increasing night-time temperatures, while CO_2-induced warming would result in an almost equally large rise in minimum and maximum temperatures;
2. higher temperatures could have a positive effect on growth of plants of the CAM type. They would also strengthen the CO_2fertilization effect and the CO_2 anti-transpirant effect of C_3 and C_4 plants unless plants get overheated;
3. higher night temperature may increase dark respiration of plants, diminishing net biomass production;
4. higher cold-season temperatures may lead to earlier ripening of annual crops, diminishing yield per crop, but would allow locally for the growth of more crops per year due to lengthening of the growing season. Winter kill of pests is likely to be reduced at high latitudes, resulting in greater crop losses and higher need for pest control;
5. higher temperatures will allow for more plant growth at high latitudes and altitudes.

Combined Effects and Some Uncertainties

The changes in CO_2 tropospheric ozone and increased UV-B do not necessarily occur simultaneously: CO_2 increase is worldwide, but with a strong seasonality in middle and higher latitudes; significant increase of UV-B is largely limited to subpolar regions (and mainly during the northern hemisphere winter months); high near-surface O_3 levels are restricted to the neighbourhood of major cities, airports, etc.

The Hydrological Cycle and Soils

Even a slight increase in surface temperatures will affect evaporation, atmospheric moisture and precipitation. While it is generally agreed that rainfall will increase (by an estimated 10 to 15%), two aspects have to be elucidated: how will rainfall intensities be affected, and what are the details of spatial changes. This is still largely a matter of discussion among experts.

Based on palaeoclimatic analogies, certain authors predict more favourable rainfall conditions in the present-day Sahel. If the increase in precipitation should be associated with increased rainfall intensities, then the quality and quantity of soil and water resources would decline, for instance through increased runoff and erosion, increased land degradation processes, and a higher frequency of floods and possibly droughts. However:

1. the extra precipitation on land, if indeed including present subhumid to semi-arid areas, will increase plant growth in these areas, leading to an improved protection of the land surface and increased rainfed agricultural production; in already humid areas the extra rainfall may, however, impair adequate crop drying and storage;
2. the extra precipitation predicted to occur in some regions provides possibilities for off-site extra storage in rivers, lakes and artificial reservoirs (on-farm or at subcatchment level) for the benefit of improved rural water supply and expanded or more intensive irrigated agriculture and inland fisheries:
3. the effects on water resources and water apportioning of international river and lake basins can be very substantial, with political overtones.

The greatest risks are often estimated to be associated with increased soil loss through erosion.

Soils, as a medium for plant growth, would be affected in several other ways:

1. increased temperatures may lead to more decomposition of soil organic matter;
2. increased plant growth due to the CO_2 fertilization effect may cause other plant nutrients such as N and P to become in short supply; however, CO_2 increase would stimulate mycorrhizal activity (making soil phosphorus more easily available), and also biological nitrogen fixation (whether or not symbiotic). Through increased root growth there would be extra weathering of the substratum, hence a fresh supply of potassium and micronutrients;
3. the CO_2 fertilization effect would produce more litter of higher C/N ratio, hence more organic matter for incorporation into the soil as humus; litter with high C/N decomposes slowly and this can act as a negative feedback on nutrient availability;
4. the 'CO_2 anti-transpirant' effect would stimulate plant growth in dryland areas, and more soil protection against erosion and lower topsoil temperatures, leading to an 'anti-desertification effect'.

Global climate change, if it occurs, will definitely affect agriculture. Most mechanisms, and two-way interactions between agriculture and climate, are known, even if not always well understood.

It is evident that the relationship between climate change and agriculture is still very much a matter of conjecture with many uncertainties; it remains largely a conundrum.

Major uncertainties affect both the Global Circulation Models (GCMs) and the response of agriculture, as illustrated by differences among models, especially as regards effects at the national and subregional levels. In addition, many of the models do not take into consideration CO_2 fertilization and improved water-use efficiency, the effect of cloud cover (on both climate and photosynthesis), or the transient nature of climate change.

It is also worth remembering that enormous knowledge gaps still affect the carbon cycle (with a missing sink of about 2 Gt of carbon), the factors behind the recent near-stabilization of the atmospheric methane concentrations or the unexplained reduced rate of CO_2 increase in recent years, the effect of volcanic eruptions (such as the recent Pinatubo eruption), the effect of any increased cloudiness, etc.

References

FAO. 1995. *World Agriculture: Towards 2010, An FAO Study*. N. Alexandratos (ed.). John Wiley, Chichester, UK, and FAO, Rome. 488 p.

Gommes, R. 1993. Current climate and population constraints on world agriculture. In: *Agricultural Dimensions of Global Climate Change.*H.M. Kaiser and T.E. Drennen (eds.). pp. 67-86.

IPCC. 1992. *Climate Change 1992.* The supplementary report to the IPCC scientific assessment. J.T. Houghton, B.A. Callandar and S.K. Varney (eds.). Cambridge University Press. Cambridge.

Kukla, G. and Karl, T.R. 1993. Nighttime warming and the greenhouse effect. *Envir. Sci. Technol.* 27 (8): 1468-1474.

Seitz, F. 1994. *Global Warming and Ozone Hole Controversies, A Challenge to Scientific Judgement.* G.C. Marshall Institute, Washington DC. 32 p.

Tans, P.P., Fung, I.Y. and Takahashi, T. 1990. Observational constraints on the global atmospheric CO_2 budget. *Science* 247: 1431-1438.

3

Climate Change and the Availability of Water Resources

The development of large water resource systems can take 10 to 20 years from the initiation of preliminary studies to realization of projects, and therefore an assessment of the impact of global warming on water resources is urgently required.

Methods for evaluating future available water are discussed and examples given of impact studies using outputs from General Circulation Models (GCMs). In this respect, Beran has emphasized a distinction between hydrology and available water resources. The former deals with evaluating the total resource (naturalized river flow) whilst the latter requires the quantification of the exploitable amount.

The world's agricultural output is heavily dependent on irrigation. Out of a global cropped area of 1 500 million ha, 16% is irrigated. Global warming is likely to have a major impact on the hydrological cycle and consequently on irrigated agriculture. This impact may be more significant than the direct effect of higher temperatures. Even though this may be the case, it is essential to view the impact of climate change in context with other major global changes. Demands on water are multi-sectorial and irrigation is already in competition for potable water, industrial, power generation, and recreational and environmental uses: most of which command a higher price. Within many Near East and African countries current water resources are almost fully exploited and supplies will have to double over the next 20 to 30 years to maintain the current, even unsatisfactory status quo. When the problem of increasing levels of water pollution is added into the equation, a situation arises of 'Water in Crises'.

Unfortunately, in most cases hydrology is treated on a national, rather than a regional, continental or global scale. It is heavily dependent on accurate data collection. With national water resource agencies generally small units within major ministries, and having low priority, funding has fallen to such an extent that in the majority of African countries reliable data are no longer collected.

The seriousness of the problem has yet to be addressed; although the World Bank is promoting a new 'Integrated Approach to Water Resource Management for Sub-Saharan Africa' and other regions of the world, whilst WMO is attempting to set-up a 'World Hydrological Cycle Observing System (WHYCOS)' and FAO is establishing a programme to develop a worldwide water database with hydrological modelling capability using Geographic Information System (GIS) and remote sensing technology.

Methods of Evaluation

Different approaches can be used to estimate the effect on water resources of global warming of which the main ones are:

1. use of instrumental records;
2. palaeoclimatic analogues;
3. GCM outputs.

In the first approach the natural variability of climate is used to predict rainfall patterns during sequences of extremes of dry or wet years. Difficulties arise due to the short period of records, although stochastic data generation techniques can be used to extend the database. Over such limited time spans, however, longer-term changes in vegetation, sea surface temperatures, ocean currents, etc., which would be expected to take place with a doubling of atmospheric forcing functions such as CO_2 are not adequately accounted for. Irrespective of this limitation, significant long-term changes in surface runoff have occurred over the past 30 years in a number of major river basins, especially in Africa. Recently, strong scientific evidence has also been produced to suggest that such changes are directly related to global warming triggered by the increase in anthropogenic greenhouse gases. As some of the recorded trends in river flows are of the same order of magnitude as would be expected for a 2 x CO_2 scenario, a review of recent changes in the hydrological cycle adds to an understanding of the problems associated with long-term changes in gross water resources (section *Recent changes in global runoff*).

Palaeoclimatic analogues fall into the same category as instrumental records though with obvious drawbacks: estimating past climates is extremely unreliable and fragmentary. Their main benefit lies in being able to cover the range of temperatures likely to be experienced over the next century or two (Table 1).

Recent studies suggest that global temperatures during previous ice ages fluctuated widely over time periods as short as decades or centuries (Heinrich cycles) and temperatures were not relatively constant as generally assumed.

Table 1. Past global temperatures

Mean global temperature above present-day values	Date
+ 1.5°C	6 000 BP
+2.5°C	125 000 BP
+ 4.0°C	4 000 000 BP

There is also no consensus amongst scientists over the prime cause for past temperature changes in the earth's atmosphere, or for the triggering of the ice ages; it certainly was not anthropogenic. The general accepted viewpoint is that changes in solar radiation received at the earth's surface are the prime instigators. However, with current research spanning many disciplines, there has been a prolifigation of new theories related to the controlling factors responsible for producing ice ages and postglacial optimums.

Current theories range from rates at which the earth's crust is formed, extent of coverage of peat bogs, release of methane from shallow seas due to formation of sea ice, rate of mountain building and resulting erosion and sediments deposited in the oceans, and cyclic changes in iceberg production affecting the north-south ocean 'conveyor belt', possibly due to instability from trapped thermal heat below glaciers, etc.

Until the casual factors involved in past climate change are known, it would be unwise to assume that they would have no influence in determining the resulting climate; an inherent assumption in the climate analogue method. Present land use, vegetation, ice cover, atmospheric composition, etc. are all likely to be so different from those experienced during previous postglacial optimums as to make a direct comparison between past and present climates of speculative value only.

GCMs are the only credible tool for predicting climate change and for providing inputs to hydrological models. Unfortunately, at the present time it is not possible to use precipitation outputs from GCMs directly as inputs to hydrological models. The outputs are not sufficiently accurate to simulate daily and monthly sequences (Table 2).

Table 2. Comparisons between observed rainfall over England and estimates from three GCMs

Source	Dec-Feb
Observed	2.5 mm/day
UK Meteorological Office GCM (Bracknell)	1.2 mm/day
National Center for Atmospheric Research GCM (Boulder, Colorado)	3.4 - 4.5 mm/day
Goddard Institute for Space Studies GCM (New York)	3.5 - 4.5 mm/day

However, the use of GCMs is the only way forward and with improved understanding and better modelling of the systems (especially ocean circulation and the land phase of the hydrological cycle) and larger computer memories, it is only a matter of time before these deficiencies will be remedied.

General Circulation Models

A GCM model typically has four components: atmospheric, land, ocean and sea ice. As the four models are interactive, production runs must include all four model components. Initially, the atmospheric model was developed to a relatively high level of sophistication, whilst the land phase and the ocean components were very simplistic.

Runoff enters into current GCM model simulations at two points. Firstly, at the boundary between the atmosphere and the land surface, where flux transfers are converted into surface runoff, and secondly at the boundary between the ocean and land, where inflow hydrographs are required as input to the ocean model. At present neither are simulated with sufficient accuracy. Kite *et al.* have demonstrated some of the problems associated with GCM modelling of the hydro-logical cycle, particularly the lack of adequate lateral transfer of water. The problem of inadequate routing of surface runoff is highlighted by studies of Semtner (1984, 1987) where a reduction in freshwater inflow to the Arctic Ocean would have a significant effect on salinity gradient and it was calculated that the outflow from the Kara and Barent seas could halve should there be a total diversion of rivers flowing to these seas. The oceans are therefore quite sensitive to freshwater inputs.

It has been demonstrated that a high degree of 'recycling' of rainfall occurs over land and if, for instance, land evaporation is much reduced or eliminated, then a significant reduction in precipitation results. In order to determine actual evapotranspiration an accurate assessment of surface runoff is required. Current GCMs do not achieve this. It is likely, therefore, that shortcomings in the modelling of the hydrological cycle are reducing the accuracy of outputs from current GCMs, as has been highlighted by Rind *et al.*, The GISS GCM, and others, have been shown to greatly underestimate actual evapotranspiration (ET) through overestimating the reduction in transpiration with increased soil moisture deficits. For example over the USA the ratio of Actual ET/Potential ET for current climatic conditions was estimated by GCM to be 0.23, whereas the actual ratios should lie between 0.6 and 0.9. Such a large discrepancy arises in part due to vegetation cover being inadequately represented in the land phase in GCMs. It was concluded by Rind that ET loss has been seriously underestimated by GCMs and that global precipitation increases of 9 to 15% are not compatible with a 30% increase in atmospheric water holding capacity corresponding to a 4°C warming. Either the rainfall estimates are underestimated or much more serious droughts will occur over continents.

Initially, climate change was seen, understandably, as a meteorological problem. The importance of the atmospheric-ocean interface was always known, but only now are results from coupled models becoming available. Similar improvements are required in the hydrological cycle components. The Global Energy and Water Cycle Experiment (GEWEX) of the World Climate Research Programme of ICSU/WMO/IOC should provide the impetus for such improvements. However, until the land phase is adequately modelled by GCMs then there is little hope that its outputs can be used with confidence in water resource planning. Furthermore, river basin impact studies undertaken to date have been based on results obtained from equilibrium GCMs in which the oceans are modelled simply as a fixed layer(s). Because of this major limitation, they are only used to simulate equilibrium states such as 1 x CO_2 or 2 x CO_2 neither states of which are, or will be, in equilibrium. All the GCMs are structurally similar and contain the same basic algorithms and will inherently include the same errors. It is rather disturbing, therefore, that such large differences in output values are produced. Equilibrium models are relatively stable and do not suffer to the same extent from cold start and drift problems that afflict transient models. Because of the shortcomings of the equilibrium models the reliability of the outputs, particularly precipitation, is seriously compromised. The range in outputs from the different equilibrium models is generally so large that it is often not possible to determine the sign of the change in runoff over large regions or even over some continents. Because of this problem some researchers have used outputs from a number of GCMs and developed levels of probability for decreases (or increases) in temperature and precipitation. Although the results from such a risk analysis approach present a useful guide for water resource planning, when all the other uncertainties are included, such studies are of little practical value in the detailed planning of future water resource development, except to make the planner aware of the increased uncertainty due to climate change.

A significant improvement in recent years has been the development of transient, (ocean-coupled) GCMs. However, the computer time utilized and the costs in running these models are at present exorbitant. As an interim measure techniques have been developed which allow for the rescaling of equilibrium model results and integrating them with those of individual transient model runs to provide an indication of temporal changes

The first transient model experiment (UKTR) undertaken by the Hadley Centre determined the climate's response for an increase in CO_2 concentration of 1% per annum over a 75 year period. The northern hemisphere was found to warm twice as rapidly as the southern hemisphere. The rise in temperature of the Antarctic was of a similar magnitude to that for the southern hemisphere as a whole, i.e., very different from previous equilibrium GCM results. In addition, the temperature rise in the transient model experiment was only 60% of that for the equilibrium model, due to the large inertia of the oceans. The change in global rainfall was, however, similar to that produced by early UK Meteorological Office GCMs. Further experiments have attempted to model historical changes in climate from 1860 to 1990 incorporating for the first time the attenuating effects of sulphate aerosol emissions. The results are encouraging, and the anomalous reversals in global temperature rise in the early 1940s, and the rapid warming from the mid-1970s, are reproduced by the model. Evidence is mounting that the recorded changes in climate over this century are consistent with those expected from increasing emissions of greenhouse gases.

Unfortunately, transient coupled GCMs are not yet capable of reproducing dynamic regional sea surface temperature patterns, and the associated precipitation changes, to the accuracy required to simulate the trends in regional precipitation which have occurred this century. To do this it would be necessary to understand the mechanisms which trigger the Southern Oscillation and to be able to simulate both them and those of the north Atlantic ocean conveyor system, and other significant ocean circulations. It is not possible, therefore, to confirm at the present time, using outputs from transient GCMs, that recorded global precipitation changes are the direct results of global warming. Until such calibrations are effected, then precipitation outputs from GCMs, transient and equilibrium, will remain speculative and should not be used directly in detailed planning of new water development projects.

Hydrological Models

The models under discussion here are concerned with the land phase of the hydrological cycle, and generate river flow and groundwater recharge. The most commonly used is the deterministic model. It is a 'conceptual' type model used in catchment hydrology to simulate at various levels of detail the physical processes involved in moisture flux transfer: from interception, evapotranspiration, soil moisture changes, surface runoff, infiltration, percolation, interflow, groundwater storage and channel routing. They tend to be lumped processes and are the workhorse of applied catchment hydrology. An essential prerequisite in applying these models is a high level of calibration based on recorded data.

Much research in pure hydrology is aimed at improving the individual processes and obtaining better spatial representation of input data. This, eventually, should reduce the importance of calibration. However, there are pitfalls in using over-sophisticated representation of the hydrological cycle in one or two processes at the expense of over-simplification in others, especially without evaluation and verification on actual catchments. Similarly, statistical and black box models, often regression based, have little use in impact studies, where many parameters affecting runoff are non-stationary, e.g., evapotranspiration, vegetation, land use, groundwater abstractions, etc.

Simple water balance models, or models incorporating evapotranspiration based solely on temperature, are also of little value either in water resources impact studies or in generalized estimates of regional changes. There is no short cut to evaluating regional change in water resources without involving all the components of the hydrological cycle at a basin or subbasin scale. Time scales in hydrology are extremely important. For example, in arid and semi-arid regions the annual runoff is dependent on rainfall intensities at an hourly time scale falling over relatively small areas and on percolation within the river bed. Times to peak for catchments of areas of thousands of square kilometres may be only one or two hours. In such catchments the intensity is as crucial to surface runoff as the rainfall amount. Similarly, if annual potential evapotranspiration were to increase by 25% over a catchment with few rainfall days in the year, it would have little impact on the volume of runoff generated. Simple annual and monthly water balance models will work satisfactorily in such conditions. In catchments with significant soil storage, the runoff is governed by available storage and rainfall intensity. In the simple balance equation:

$$Q = P - E$$

where

Q = surface runoff,

P = precipitation,

E = catchment losses.

Q is not independent of E, and if treated as such, will produce significant distortion when calculating changes in Q relative to changing of P and E.

Also, in most catchments both geology and vegetation cover can be extremely variable. These parameters alone produce large differences in seasonal and annual runoff. For example in the River Canje catchment in Guyana, half the catchment is covered by tropical forest with a closed canopy and underlain by pervious 'White Sands'; whilst the remainder is largely covered by savanna and relatively impermeable sediments. As a result, runoff ratios (Q/P) in the wet season are about 0.15 in the forested area, and 0.85 in the savanna; in the dry season the values are 0.50 and 0.10, respectively

Interfacing GCMs and Hydrological Models

There are many ways of using outputs from GCMs to provide inputs into hydrological models. The most obvious is to take the GCM simulations of daily rainfall, temperature and other meteorological variables and feed them directly into a calibrated hydrological model. Unfortunately precipitation outputs from different equilibrium models are extremely variable. Also, because of the way clouds and precipitation are modelled, GCMs produce unrealistic simulations of day-to-day rainfall. An alternative approach is to use the 'change' in rainfall, temperature and other climate variables either to produce directly a perturbed time series or to produce stochastically generated sequences of wet or dry states with amounts based on conditional probability distributions. This latter method is a compromise to cover for the inadequacies of the GCM outputs and to produce more realistic looking data sequences. Similarly, regional GCM-generated pressure patterns can be used with a suitable regional multivariate model to generate sequences of rainfall and evapotranspiration. However, the pressure patterns cannot be used directly to produce climatic sequences. None of the methods used gives confidence that results will be other than approximate and of limited value in water resource planning.

Impact Studies of Climate Change

To date, most of the impact studies undertaken have not used hydrological models which adequately simulate the transient nature of the parameters of the hydrological cycle. Even fewer have attempted to include changes in water demand.

Many impact studies have found that changes in evapotranspiration rates due to global warming have only a marginal effect on surface runoff when compared with changes in precipitation. These conclusions have been questioned. Rind *et al.* have suggested that most GCMs fail to simulate the land phase satisfactorily and underestimate actual evapotranspiration rates. The effect of increased potential evapotranspiration may significantly reduce surface runoff in the tropics. In addition, often changes in temperature alone have been used to estimate changes in evapotranspiration. This again will underestimate actual evapotranspiration. Finally, many studies have used simplified hydrological models, often of the water balance type, to analyse changes in surface runoff. Such methods can easily be demonstrated to be inappropriate.

The Institute of Hydrology (IH) has undertaken an impact study for 21 catchments in England and Wales. The wettest scenario produces a general increase in average runoff across England and Wales, whilst the driest produces a reduction. However, the most interesting findings relate to changes in evapotranspiration. Three different potential evapotranspiration (PET) scenarios were investigated: PET changes based on; Case 1 - increased temperatures only; Case 2 - changes in temperature, net radiation, relative humidity and windspeed; Case 3 - Case 2 plus changes in plant stomatal conductance and leaf area. The results are given in Table 3.

Table 3. Equilibrium climate change scenarios for 2050

Climate parameter	Annual changes in England and Wales
Temperature	+2.2%
Rainfall	+4%
PET 1	+10%
PET 2	+35%
PET 3	+ 24%

An important feature of the results is the quite dramatic increase in PET predicted in England and Wales resulting from the inclusion of climatic elements other than temperature. A change in humidity produced the most significant increase based on PET calculations using the Penman-Monteith formula.

Kite has recognized the importance of land cover changes in water resource impact studies. The Kootenay catchment in Canada was divided into 50 Grouped Response Units (GRUs) and 10 land cover divisions, giving a potential for subdivision to 500 units. The land cover data were obtained from satellite data. Good calibration was obtained using a distributed hydrological model (SLURP). In the 2 x CO_2scenario with changed land cover, the frequency of high flows was found to more than double and evapotranspiration was found, surprisingly, to decrease by 10%.

Over the last century the hydrological cycle over most river basins has been subjected to extensive human changes with the construction of reservoirs, land-use changes, river abstractions, groundwater abstractions, inter-basin diversions, etc. Such changes significantly alter river flows and the hydrology of the basin. For example, the Nile loses 10% of its flow through reservoir evaporation above the Aswan Dam whilst the Zambezi loses the equivalent of almost 20% of its flow at Victoria Falls from evaporation losses from the Kariba reservoir. On a global scale it has been estimated that sea levels would have risen this century by 2 mm per year, rather than the recorded 1 mm per year, had it not been for increases in reservoir storage and river abstractions, mainly for irrigation purposes. It is important, therefore, that existing modifications to natural river flows are evaluated in impact studies as well as likely future changes in demand and supply.

It is evident that forecasts of future water availability produced by climate impact studies to date have limited value in present-day detailed water development designs. Their main value lies in improving impact modelling techniques. Certainly, too much effort has been placed on routine impact studies at the expense of targeted research. Estimates of the effect of climate change on precipitation and evapotranspiration are not sufficiently accurate to provide credible hydrological predictions.

Recent Changes in Global Precipitation

Hulme *et al.* have prepared maps showing global precipitation as well as annual and seasonal changes in precipitation between the periods 1931 to 1960 and 1961 to 1990. The two most prominent features are over northern Russia with increased precipitation of up to 20% and the African Sahel, between latitudes 10°N and 30°N, with reduced precipitation between 20 and 50%. Significant changes have occurred over all continents. These recorded changes over the last 60 years are of a similar magnitude to those predicted by GCMs for a 2 x CO_2. How far these recorded changes are related to radiative forcing due to increased greenhouse gases is controversial. The standard response is to state that the natural variability in climate will mask any forcing for two or more decades. This viewpoint can now be questioned following the results of several recent research studies, most notably Thomson. In a major study of annual solar variations at the earth's surface, it has been demonstrated that, due to approximations in using the tropical year (equinox to equinox) in the Gregorian calendar rather than the anomalistic year (perihelion to perihelion), the starting dates of the seasons should be occurring later each year this century. A reversal occurred in 1923 which cannot be explained by changes in incoming solar radiation or by other phenomena such as the occurrence of volcanic eruptions or the Southern Oscillation, but is completely consistent with recorded increases in greenhouse gas emissions. Moreover, were the recorded increases in global temperatures due to greater solar radiation activity, then the amplitudes in seasonal temperature would be widening. The reverse is the case, which is consistent with the greenhouse effect. In addition, UK Meteorological Office GCM modelling of historical changes of climate (1960-1990) has reproduced both the anomalous reversals in global temperature and the rapid increases recorded later in the century. The evidence is mounting, therefore, that recorded changes in climate this century are the direct result of increased greenhouse gas emissions. If this is the case, then changes in river flow regime over this century probably present the best evidence available for water resource planning of the transient effect of global warming.

Although there are no transient GCM results to make comparisons between the observed precipitation change recorded between 1931 to 1960 and 1961 to 1990, an interesting comparison can be made between present-day conditions and those predicted in 50 years' time. On a global scale the same pattern of change is modelled. The significant exception is that of Africa, north of the equator, which is shown to experience increased rainfall. The Sahelian drought may therefore be ameliorated to some degree. This is consistent with a predicted reversal in hemispheric warming with the north warming much more rapidly than the south; the opposite situation to that seen over the past 30 years. Aridity is predicted to increase in Australia, a trend which has been apparent in recent years, although clearly associated with the Southern Oscillation, a feature not represented at a decadal scale in model outputs. The other continents are shown to continue the trend of the past 30 years. However, considerable reservation must be placed on such GCM predictions until the changing rainfall/runoff patterns experienced over the past 100 years have been simulated with a measure of accuracy. This will require a much improved understanding of ocean circulation.

Factors Affecting African Hydrology

The Hurst Effect

Hurst showed that annual Nile floods did not follow the classical theory of a stationary stochastic process. The estimated storage to maximize yield of the Nile was much higher than classical theory indicated, e.g., for a 500-

year period Hurst calculated an accumulated departure of flow from the River Nile of 48 standard deviations from the mean compared with the theoretical 28. This problem was a dominant feature of hydrological research in the 1970s. Klemês, in a comprehensive review of research into the Hurst phenomenon, warned of the dangers of deluding ourselves that statistical models such as fractional Brownian noise with its implicit acceptance of infinite memory could be used to explain the phenomenon. A highly successful operational model may turn out to be totally unacceptable from a physical point of view - the Ptolemaic planetary model for example. Klemês also demonstrated that the Hurst phenomenon could be produced using a zero memory process and non-stationary means. Klemês even suggested that operational models of the future may revert to extrapolations based on sound short-term knowledge rather than long-term synthetic data based on 'ignorance and guesses'.

Since the 1970s, interest has waned in trying to explain mathematically why so many long-term natural series fit distributions which are not Gaussian. As the modelling of physical processes has developed, the mechanisms which can lead to persistence of high and low rainfall have become better understood.

Physical Feedback Processes

The climate of Africa is rather unique. In comparison with the other continents, Africa south of the Atlas Mountains has no major continuous mountain barriers, either longitudinally or latitudinally orientated, to disrupt the circulation. Consequently, its climates and its seasonal variations are less complex and governed to a large extent by the oscillating inter-tropical front which follows the sun on its annual passage between the two hemispheres.

Nicholson has clearly demonstrated, in a reconstruction of sequences of historical droughts and wet periods of Africa, that the African climate bridges the two hemispheres. The extreme drought years of 1972 and 1984 which devastated the Sahel region from the Atlantic coast to Ethiopia are clearly reflected in the flows of the Rivers Nile, Chari, Zaire, Niger, Zambezi and Orange in these years. Although a high degree of homogeneity does exist between climatic fluctuations in the north and south of the continent, the influences of such large-scale systems as the Southern Oscillation and the Atlantic Oscillation, both in part driven by sea temperature variations, and the monsoons of Asia, do produce spatial anomalies in climate extremes in different areas of Africa.

There is a tendency for an equatorial band across African to react in an opposite direction in terms of precipitation to the rest of Africa. This has been demonstrated clearly by the recent high levels recorded in Lake Victoria, Lake Tanganyika and Lake Malawi. Perhaps the clearest example of this anomaly is that the higher White Nile flows, originating in equatorial Africa, have persisted from 1961, whilst in contrast the Blue Nile has suffered the most serious reductions in discharge recorded this century, or over the past 1 300 years if the Roda Nilometer records are accepted as a reliable source. It seems that for extreme wet and dry years the whole continent can bc affected with only minor deviations. Nicholson isolated four different preferential climatic patterns over the African continent. These modes are:

— northern and southern subtropical areas dry; equatorial regions wet;
— northern and southern subtropical areas wet; equatorial regions dry;
— whole continent dry;
— whole continent wet.

Dry continent and dry subtropical area modes have dominated the continent since 1970.

From a water resource planning viewpoint the major concern is whether or not the drought, which started in the late 1960s in the Sahel, and became dominant in southern Africa throughout the 1980s, will continue. One viewpoint, which has been strongly supported in the past in southern Africa, is the existence of cyclic behaviour. Such fluctuations, and particularly the persistence in climate whereby one dry year seems to be followed by another and one wet year by another wet year, can be explained by biogeophysical short-term feedback processes, the so called 'Joseph' and 'Noah' effects: reduced vegetation leads to increased albedos and increased radiation losses, surface cooling and greater atmospheric stability which reduces rainfall and encourages persistence. Similar persistences are produced by lowering soil moisture levels in GCM simulations, indicating the importance of rainfall itself in initiating a significant feedback process, and by the increase in atmospheric particles from sandstorms. Such feedback processes have always been around and explain the Hurst effect and almost certainly explain much of the natural persistence of sequences of wet years and dry years found in the Sahel region, although fluctuations in sea surface temperature also play a role.

Water resource engineers in the past have relied on statistical data, whether stochastically generated or historic, for their estimation of reliability of yield. Long-term trends are generally ignored unless physical explanations can be presented to justify their inclusion. In this respect the Sahelian drought has been with us for almost 30 years and suggests that it has been maintained by causal factors other than normal feedback process. Global warming is an obvious explanation.

Global Causes

The search for a causal explanation of the African drought received an impetus by research undertaken in the UK Meteorological Office, which demonstrated a strong correlation between sea surface temperature (sst) anomalies on a global scale with wet and dry periods in the Sahel. After the mid-1960s, a marked cooling of the oceans of the northern hemisphere and simultaneous warming in the southern hemisphere was observed. A reversal occurred around 1970, since when temperatures in both hemispheres have increased. A time series plot of sst differences between oceans of the southern hemisphere and those of the northern hemisphere, and rainfall anomalies for the Sahel, shows a strong negative correlation. The correlation between the July-September sst and Sahel rainfall for the period 1901 to 1984 was -0.62, which is significant at the 99.9% probability level. Numerical equilibrium GCM experiments with prescribed sea temperatures were also undertaken by the UK Meteorological Office which were able to replicate rainfall reductions in the Sahel for recent drought years: 30% reductions over western Sahel, 20% over eastern Sahel and up to 50% over the mountains of southern Sudan and northeast Ethiopia.

The observed and unexpected warming of the southern oceans at a faster rate was thought to be due to a reduction in the heat transfer from southern to northern hemispheres, although the detailed mechanisms of the transfer are still the subject of much research. Alternative scenarios include increased deep water circulation from 1960 to 1970 in the Atlantic and the effect of sulphate aerosols which are dominant the northern hemisphere. If, however, the reduction in heat transfer is related to the north-south conveyor system combined with a slowdown in formation of north Atlantic deep water at high latitudes, owing to a reduction in the extent of sea ice, then the Sahelian drought may persist until the greater land mass in the northern hemisphere starts to dominate the effects of a slowdown in ocean transfer and the attenuation effect of sulphate aerosols. However, confirmation will depend on further research developments into detailed coupled transient GCM models, which can be calibrated against recent climates and sea temperatures. It is likely that the results from the equilibrium GCMs will be found wanting once more reliable coupled models have been developed.

A plausible scenario of future global warming links a weakening in carbon sinks and radiation sinks in the polar regions with reduced deep water formation due to reduced heat transfers from the southern hemisphere to the north. It is postulated that reductions to the radiational and CO_2 sinks could give rise to significant positive feedbacks leading to an increase in global warming.

Recent Changes in Global Runoff

Lake Chad

Lake Chad is situated on the border of the Sahara between latitudes 12°N and 14°30'N and longitudes 13°E and 15°30'E. The lake has no outlet to the sea and is the main focus of an internal drainage basin of 2 500 000 km^2 that collects water from Algeria, Chad, Niger, Nigeria, Cameroon, Sudan and the Central African Republic. Over 90% of the total lake inflow comes from the Chari-Logone river system which rises on the southern margin of the basin and has a combined catchment of about 570 000 km^2. Lake Chad is located centrally within the Sahel, which is often defined as the 250 to 500 mm rainfall band across Africa below the Sahara desert. Lake Chad is shallow; average depths vary between 1.5 and 5 m, with the result that its surface area is very sensitive to changes in inflows. The lake has a number of unusual features. In spite of being an enclosed basin, it remains a freshwater lake. Its size ranges from 100 km^2 for the 'Gran Chad' to 10 to 30 $km^2$for the 'Petit Chad'. The water tends to be replaced every year or every other year, with an average annual inflow of about 30 to 40 km^3. The saline water in normal times is pushed to the northwest of the lake where its salinity is concentrated and natural seepage to the north maintains the salt balance in equilibrium.

Feasibility studies began in 1972 to investigate the irrigation of lands bordering Lake Chad in Nigeria. As a result, the South Chad Irrigation Project (SCIP) was designed for three-stage implementation. Two stages have been built: Stage I in 1979 and Stage II in 1983. A pump station with a total capacity of 103 m^3/s (current capacity 75 m^3/s) was constructed to pump water from the lake via a canal. The full scheme has a gross area of 66 000 ha and would abstract approximately 3% of the annual inflow to Lake Chad. The layout of the scheme was based on 4 ha units to provide 16 000 households, or a 100 000 population, with a more secure livelihood. It was planned to produce two crops per year based on rice (50%), cotton (20%) and wheat (60%), with a cropping intensity of 130%.

In the mid-1960s Lake Chad had reached record levels for this century giving rise to speculation that the 'Gran Chad' as observed by explorers of the last century would be re-established. However, this failed to happen and lake levels fell during the 1970s and at the time that the feasibility studies were completed were similar to those experienced during the first half of the century. No one anticipated the drastic fall in rainfall which has occurred since.

Table 4. River Chari annual runoff (mean flow before 1971 - 40 km^3/annum)

Year	Flow	Year	Flow	Year	Flow
1971	31	1978	29	1986	15
1972	18	1979	21	1987	10
1973	18	1980	25	1988	27
1974	29	1981	19	1989	16
1975	35	1982	21	1990	12
1976	29	1983	17	1991	19
1977	25	1984	7.5	1992	21
		1985	17	Mean annual flow	22

The sudden onset of the Sahelian drought in the 1970s resulted in a severe contraction in the size of the lake, changing its character to a swampy delta at the mouth of the River Chari, with dire consequences for lake-side communities and the SCIP irrigation scheme.

Lake Chad divided into northern and southern pools in 1973 and by 1976 levels had fallen below the bed of the intake canal. Although water levels recovered during October and November, providing irrigation water for wheat crops planted in November, shortage of water in June to September prevented the cultivation of the main rice crop. The two pools have remained separated and since 1983 levels have continued to fall and have prevented even winter-season irrigation. The scheme is maintained in good condition awaiting the recovery of the lake.

The reason for the fall in lake levels is simply that flows in the River Chari-Lagone basin failed. The total runoff in 1984 was only around 20% of the long-term mean. Since 1971 flows have been reduced to 50% of the previous long-term mean annual flows of over 40 km^3, although rainfall is estimated to have fallen by only 25%. Of all the major basins in the world, probably Lake Chad has been affected most by climate change. The most ambitious irrigation scheme built in Africa since the mid-1940s has yet to be operated in earnest

Falling Lake Chad levels demonstrate the effect of a rainfall reduction of 25% on river flow in a semi-arid region where rainfall gradients are steep and small shifts in the atmospheric circulation can cause large rainfall changes. As a result, proposals have been put forward for a major inter-basin transfer from the River Zaire to the River Chari basin. With an annual runoff of 1 250 km^3, potential supplies are available.

River Nile

To a large extent Egypt has remained unscathed by the drought owing to the large over-year storage in Lake Nasser. However, other factors have contributed to reducing the drought's impact. One of these is the exceptionally high levels in Lake Victoria which have helped maintain higher White Nile flows. The higher levels have resulted from very heavy rainfall in Kenya and Uganda between 1961 and 1963 and above average rainfall since. The higher White Nile flows have helped to compensate for the lower discharges in the Blue Nile, which have been more adversely affected by the Sahelian drought.

The construction of the High Aswan Dam in 1963 provided the complete regulation of the annual Nile flood and also gave sufficient over-year storage to supply Egypt's share under the Nile Waters Agreement of 55.5 km^3, with a very high level of reliability (96%).

Following the failure of Nile flows in 1984 (a naturalized inflow to Lake Nasser of 59 km^3 was recorded, compared with a mean flow of 84 km^3), there was considerable concern that Lake Nasser's resources would fall to a level where irrigation supplies would have to be reduced with both agricultural and political implications.

Effect of Sahel drought on Nile flows

A main feature of the Nile flow series, apart from the high flow period at the end of the last century, is the steep fall in discharge from the mid-1960s. It is more pronounced and persistent than any previous low flow periods. An analysis of Roda Nilometer data also indicates that during the last two centuries the variability of the annual flood far exceeds that observed since records began in AD 622.

Most previous yield studies of the River Nile have analysed River Nile flows at Aswan. However, the flow records over the past two decades have demonstrated the importance of the different rainfall regimes over the catchments of the Blue and White Nile. Flows in the White Nile between 1962 and 1985 have increased by 32% or 8 km^3 above the 1912 to 1961 mean. This has occurred at a time when Blue Nile flows have decreased by 9 km^3 over the period 1965 to 1986, or 16% below their 1912 to 1964 mean. It could be inferred that the two rainfall regimes are negatively correlated. However, past records show that the relationship between annual inflow to Lake Victoria and recorded Blue Nile flows at Khartoum is random. This, however, contradicts results of GCM outputs which suggest that rainfall over the White and Blue Nile catchments may be negatively correlated.

Lake Nasser simulations

River Nile flow sequences were used to investigate the ability of the Nile systems to withstand drought should the Sahelian drought prove to be a persistent feature of climate of the region in the coming decades. Two sequences were prepared for severe and moderate droughts (based on flow statistics from 1968 to 1988). They indicate that Lake Nasser releases would have to be reduced below 55.5 km^3 (Egypt's allocation under the 1959 Nile Waters Agreement) to 52 km^3 and 49 km^3, respectively, and that these yields would have reliabilities of 87% and 88%, i.e., water shortages would occur on average in 13 and 12 years respectively every 100 years. The effect of the Sahelian drought on River Nile flows and irrigation in Egypt, although problematical, would be nowhere near as severe as for Lake Chad. Even under the most severe drought conditions the yield from Lake Nasser is predicted to fall only from 55.5 km^3 design yield to 48.4 km^3, or less than

15%, although the reliability of supply would fall from a design yield of 96 to 88%.

Options do remain, however, for increasing supplies by reducing drainage outflow and operational losses to the sea, but these will largely be needed to meet increased municipal and industrial demands. Conway *et al.*, in a recent study on future water availability in the Nile basin, utilized GCM outputs and modelled runoff incorporating land cover changes. The precipitation changes ranged from +18 mm to +39 mm and Nile flows were calculated to range from a 3% decrease to a 10% increase. These changes are much smaller that those recorded over the last two decades.

The Nile basin is a closed system and in irrigation terms is very efficient (>70%). Improvements in efficiently are therefore limited. Because it is a closed system increased groundwater abstractions do not represent extra water in the long term and provide only short-term temporary benefits. There is potential to save water by selecting less water demanding crops and the use of short-term varieties (5% savings). However, significant enhanced supplies could be obtained from the Sudd and other wetlands in Sudan but these all have major political or ecological constraints. Ignoring the political and ecological implications there is still scope for further water development within the Nile basin. However, when these are set against the population explosion in the region and possible problems of climatic change, the outlook cannot be viewed with optimism.

Caspian Sea Levels

Another region which has experienced a major change in water resources is in Russia. The River Volga, which rises near the Baltic Sea and discharges into the Caspian Sea, is the largest river in Europe, 3 530 km long. Its catchment area of 1.5 million km^2 makes it one of the major rivers in the world. Its river basin is larger than that of the Zambezi, the Indus or the Ganges. Around 60% of its discharge is derived from melting snow. Three main rivers, the Volga, Ural and Terek, account for 88% of the total river inflow to the Caspian Sea and, of this, over 50% arises from the Volga river. It is connected by canals to Moscow, Volgograd, St. Petersburg and the Arctic Ocean for river traffic and, with hydropower stations and extensive irrigation systems around the north Caspian Sea coast and the Azov Sea, the Caspian basin is crucial to the economy of Russia, Azerbaijan, Kazakhstan and the Ukraine.

The Caspian Sea has been called the greatest salt lake in the world, although its salt content is only one-third of that of the main oceans. It has no natural outlet. It covers an area of 372 000 km^2 with its surface at a level of -28.5 m below mean sea level, whilst the total basin area extends to some 3.7 million km^2. An extensive sedimentary plain surrounds the outlet of the River Volga where the sea is shallow with depths of only 4-6 m. The total capacity of the sea is 66 960 km^3, compared with an annual inflow of around 310 km^3. The annual volume of inflow would raise levels by 0.8 m if evaporation is ignored. Being an enclosed sea, levels fluctuate with inflow and historically have shown a wide variation. Also, developments within the Caspian Sea basin have seriously reduced River Volga flows. Levels fell dramatically from 1930 reaching a low in 1977, although surprisingly since this date levels have started to rise and plans are in hand to mitigate flood damage. However, without predictions of future sea levels and a knowledge of the processes causing the fluctuations, attempts to mitigate the damages resulting from too high or too low levels are likely to be both ineffective and uneconomic.

Water development

The River Volga plays a vital role in the economic life of the states through which it passes. It is used for:

- navigation (151 000 km of waterways connecting the Arctic Ocean to the Caspian Sea, with the Moscow canal, the Volga-Don canal and the Volga-Baltic waterway all with a 4.5 m navigation depth);
- power generation (a cascade of 11 hydroelectric schemes with a capacity of 11000 MW);
- irrigation (in the Caspian-Azov basins some 7.5 million ha of irrigation have been developed; 4 million ha using sprinkler systems).

Relative water consumption over the years has been compiled and forecasts made for the future based on the construction of the Volga-Don and Volga-Chograi irrigation canals.

For withdrawals of 37 km^3 and a residual flow of 266 km^3/year the Caspian Sea was in equilibrium at a level around -28.5 m amsl. This is considered to be about the optimum level for the valuable fisheries. Over the last millennium the highest level was thought to be -22 m recorded in the 1600s, and the lowest levels, -31 m in the 1400s, and -29.0 in 1977, a range of some 9 m.

Because of the extensive and dramatic water developments undertaken by the former USSR in the Volga basin since the 1930s, the first step in any evaluation is to naturalize the flow to remove the influences of human activities. But for the massive water developments, the present-day levels of 27.7 m amsl would now have returned to the 1929 high of -26.2 m amsl. The larger abstractions from the Volga river are therefore contributing significantly to reducing sea levels, albeit fortuitously, and so preventing more serious damage to irrigation areas and urban development adjacent to the Caspian Sea (most of which have been developed in recent years). There has been therefore a dramatic reversal from the problems of falling sea levels to one of rising levels and how to contain increasing flood waters.

Table 5. Abstractions from Volga/Caspian Sea, post-1930 (km^3/yr)

Water consumption	1980	1990	2000	2010
Water supply	5.6	7.3	9	10.5
Irrigation	18.6	23	33	36
Reservoir evaporation	6	7	8	9
Filling of reservoirs	3	2	0	0
Inflow to Kara-Bogaz-Gol	0	2	2	2
Water consumption in Iran	3.5	4	5	6
TOTAL	**37**	**44**	**57**	**63**

Between 1931-1960 and 1961-1990 rainfall on the Volga catchment has increased by over 10% whilst naturalized inflow to the Caspian Sea has increased by a similar amount.

The Intergovernmental Panel on Climate Change suggests increases in precipitation of between 100 and 200 mm over the winter period for a doubling in greenhouse gases. If this prediction proves correct then it would produce a very significant increase in surface runoff in the River Volga which is predominantly snow-melt fed. The additional precipitation will fall mainly as snow on frozen ground. There will therefore be little increase in losses and most of the additional snow will be converted to surface runoff and will increase River Volga discharges substantially. River Volga flows could increase by between 25 and 50%.

If the present change in climate with increased precipitation over Russia continues over the coming decades, serious problems in terms of drainage and flooding of low-lying irrigation lands are likely to be experienced. However, in the longer term the additional water would be extremely beneficial for agricultural development in the Caspian Sea sedimentary basin where 100 000 km^2 of land lies below sea level. Alternatively, surplus water could be used to replenish all of the damage inflicted on the Aral Sea: an annual flow of around 50 km^2 would be required to restore the Aral Sea to its pre-1960 condition. A surplus of 100 km^3/year of freshwater would certainly find customers in the Near East, although the task of transferring it would be formidable.

References

Arnell, N.W. and Reynard, N.S. 1993. *Impacts of Climate Change on River Flow Regimes in the United Kingdom.* Institute of Hydrology, NERC, Water Directorate, DOE, July 1993.130 p.

Hadley Centre. 1992. *Transient Climate Change Experiment.* Hadley Centre, UK Met. Office, Aug. 1992. 20 p.

Tyson, P.D. 1987. *Climatic Change and Variability in Southern Africa.* Oxford Univ. Press, Captetown. 220 p.

World Bank/UNDP/ADB/EC/French Government. 1993. *Sub-Saharan Hydro-logical Assessment.* July 1993. Washington DC. 17 p.

4

Effects of Climate Change on Soil Fertility

The main potential changes in soil-forming factors (forcing variables) directly resulting from global change would be in organic matter supply from biomass, soil temperature regime and soil hydrology, the latter because of shifts in rainfall zones as well as changes in potential evapotranspiration.

The biggest single change in soils expected as a result of these postulated forcing changes would be a gradual improvement in fertility and physical conditions of soils in humid and subhumid climates. Another major change would be the poleward retreat of the permafrost boundary, discussed by Goryachkin and Targulian. Other widespread changes would be in degree rather than in kind.

The changes in temperature but particularly in rainfall to be expected as a result of global warming are subject to major uncertainties for several reasons. Different global circulation models do not lead to mutually consistent results, and they are not yet adequately verified. Also, the interaction with changes in location and intensity of major ocean currents and resultant possible modifications in sea surface temperatures is still most uncertain, as well as the interaction with possible major changes in cloudiness and land cover and the resulting changes in albedo and actual evapotranspiration.

Indirect effects of climate change on soils through CO_2-induced increases in growth rates or water-use efficiencies, through sea-level rise, through climate-induced decrease or increase in vegetative cover, or a change in human influence on soils because of the changes in options for the farmer, for example, may well each be greater than direct effects on soils of higher temperatures or greater rainfall variability and larger or smaller rainfall totals.

Possible Changes in Forcing Variables

With these caveats, one could stipulate the following changes in forcing variables as likely to materialize sometime during the next century:

1. A gradual, continuing rise in atmospheric CO_2 concentration entailing increased photosynthetic rates and water-use efficiencies of vegetation and crops, hence increases in organic matter supplies to soils.
2. Minor increases in soil temperatures in the tropics and subtropics; moderate increases and extended periods in which soils are warm enough for microbial activity (warmer than about 5°C) in temperate and cold climates, parallel to the changes in air temperatures and vegetation zones as summarized by Emanuel *et al.*
3. Minor increases in evapotranspiration in the tropics to major increases in high latitudes caused both by temperature increase and by extension of the growing period.
4. Increases in amount and in variability of rainfall in the tropics; possible decrease in rainfall in a band in the subtropics poleward of the present deserts; minor increases in amount and variability in temperate and cold regions. Peak rainfall intensities could increase in several regions.
5. A gradual sea-level rise causing deeper and longer inundation in river and estuary basins and on levee backslopes, and brackish-water inundation leading to encroachment of vegetation that accumulates pyrite in soils near the coast.

Effects of Higher CO_2 on Soil Fertility

Higher atmospheric CO_2 concentration, increases growth rates and water-use efficiency of crops and natural vegetation in so far as other factors do not become limiting. The higher temperature optima of some plants under increased CO_2would tend to counteract adverse effects of temperature rise, such as increased nighttime respiration. The shortened growth cycle of a given species because of higher CO_2 and temperature would be compensated for in natural vegetation by adjustments in species composition or dominance. In agro-ecosystems the choice of longer-duration cultivars or changes in cropping pattern could eliminate unproductive periods that might arise because of the shorter growth cycle of the main crop.

There will be adequate time to adjust to the changes since these are expected to occur over decades, rather than years or days as in all present experimental situations.

The increased productivity is generally accompanied by more litter or crop residues, a greater total root mass and root exudation, increased mycorrhizal colonization and activity of other rhizosphere or soil micro-organisms, including symbiotic and root-zone N, fixers. The latter would have a positive effect on N supply to crops or vegetation. The increased microbial and root activity in the soil would entail higher CO_2 partial pressure in soil air and CO_2 activity in soil water, hence increased rates of plant nutrient release (e.g., K, Mg, micronutrients) from weathering of soil minerals. Similarly, the mycorrhizal activity would lead to better phosphate uptake. These effects would be in synergy with better nutrient uptake by the more intensive root system due to higher atmospheric CO_2 concentration. There is no *a priori* reason why the degree of synchrony between nutrient release and demand by crops or natural vegetation would be subject to major changes under high CO_2 conditions.

The greater microbial activity tends to increase the quantity of plant nutrients cycling through soil organisms. The increased production of root material (at similar temperatures) tends to raise soil organic matter content, which also entails the temporary immobilization and cycling of greater quantities of plant nutrients in the soil. Higher C/N ratios in litter, reported by some workers under high CO_2 conditions, would entail slower decomposition and slower remobilization of the plant nutrients from the litter and uptake by the root mat, and would provide more time for incorporation into the soil by earthworms, termites, etc. Higher soil temperatures would counteract increases in 'stable' soil organic matter content but would further stimulate microbial activity.

In all experimental situations, whether chamber-type or free-air enrichment, CO_2 increases are rapid or sudden, often to double ambient concentration, sometimes higher. The consequently rapid increases in soil organic matter dynamics and soil micro-organisms may cause temporary competition for plant nutrients. These temporary effects have on occasion been reported as negative factors affecting plant response to elevated CO_2 However, increased organic matter dynamics and microbial activity in soils are positive for the soil-plant system when CO_2 concentrations rise gradually over decades, as currently and in the recent past. Future experiments could be set up to compensate for the temporary effects caused by the suddenness of the CO_2 increase, for example by artificially higher soil organic matter contents estimated to be near equilibrium with each stepwise higher CO_2 concentration, in a range between 350 and 600 ppm.

Increased microbial activity due to higher CO_2 concentration and temperature produces greater amounts of polysaccharides and other soil stabilizers. Increases in litter or crop residues, root mass and organic matter content tend to stimulate the activity of soil macrofauna, including earthworms, with consequently improved infiltration rate and bypass flow by the greater number of stable biopores. The greater stability and the faster infiltration increase the resilience of the soil against water erosion and consequent loss of soil fertility. Thc incrcased proportion of bypass flow also decreases the nutrient loss by leaching during periods with excess rainfall. This refers to the available nutrients in the soil, including well-incorporated fertilizers or manure, but not to fertilizers broadcast on the soil surface. These are subject to loss by runoff or leaching.

These changes increase the resilience of the soil against physical degradation and nutrient loss by increased intensity, seasonality or variability of rainfall, as well as against some of the unfavourable changes in rate or direction of soil-forming processes discussed in the next sections.

If the partial pressure of CO_2 in the soil air would rise, and that of O_2 decrease to levels impairing root function, part of the benefits indicated would not materialize. The improved gas exchange with the atmosphere through increased numbers of stable biopores would tend to keep CO_2 and O_2 in the soil at 'safe' levels, at least in naturally or artificially well-drained soils. Wetland crops such as rice or jute have their own gas exchange mechanisms and would not be affected; neither would natural wetland vegetation.

The positive effect on weathering rate and plant nutrient availability would occur in soils with significant amounts of weatherable minerals, not in very deeply and strongly weathered or otherwise very poor soils.

Effects of Rainfall and Temperature Changes

In the humid tropics and monsoon climates, increased intensities of rainfall events and increased rainfall totals would increase leaching rates in well-drained soils with high infiltration rates, and would cause temporary flooding or water-saturation, hence reduced organic matter decomposition, in many soils in level or depressional sites. This may affect a significant proportion of especially the better soils in Sub-Saharan Africa, for example. They would also give rise to greater amounts and frequency of runoff on soils in sloping terrain, with sedimentation downslope and, worse, downstream. Locally, there would be increased chances of mass movement in the form of landslides or mudflows in certain soft sedimentary materials, discussed below. Soils most resilient against such changes would have adequate cation exchange capacity and anion sorption to minimize nutrient loss during leaching flows, and have a high structural stability and a strongly heterogeneous system of continuous macropores to maximize infiltration and rapid bypass flow through the soil during high-intensity rainfall.

In subtropical and other subhumid or semi-arid areas, the increased productivity and water-use efficiency due to higher CO_2 would tend to increase ground cover, counteracting the effects of higher temperatures. If there would be locally much less rainfall and increasing intra- and inter-annual variability, these could lead to less dry-matter production and hence, in due course, lower soil organic matter contents. Periodic leaching during high-intensity rainfall with less standing vegetation could desalinize some soils in well-drained sites, cause increased runoff in others, and lead to soil salinization in depressional sites or where the groundwater table is high. Soils most resilient against the effects of such increasing aridity and rainfall variability would have a high structural stability and a strongly heterogeneous system of continuous macropores (the same as in the tropics); hence a rapid infiltration rate, as well as a large available water capacity and a deep groundwater table.

Higher temperatures, particularly in arid conditions, entail a higher evaporative demand. Where there is sufficient soil moisture, for example in irrigated areas, this could lead to soil salinization if land or farm water management, or irrigation scheduling or drainage are inadequate. On the other hand, recent experiments by the Salinity Laboratory, Riverside, California, point to increased salt tolerance of crops under high atmospheric CO_2.

In temperate climates, minor increases in rainfall totals would be expected to be largely taken up by increased evapotranspiration of vegetation or crops at the expected higher temperatures, so that net hydrologic or chemical effects on the soils might be small. The negative effect on soil organic matter contents of a temperature rise might be more than compensated by the greater organic matter supply from vegetation or crops growing more vigorously because of the higher photosynthesis, the greater potential evapotranspiration and the higher water-use efficiency in a high-CO_2 atmosphere. The temperate zone would thus be likely to have the smallest changes in soils, even in poorly buffered ones, directly caused by the effects of global change. A minor and probably slow, but very visible, change could be a reddening of presently brown soils where increased periods with high summer temperatures would coincide with dry conditions, so that the iron oxide haematite would be stable over the presently dominant goethite. This mineralogical change might decrease the intensity and amount of phosphate fixation. An overview of such changes, with emphasis on temperate climate zones, is given by Buol *et al.*

In boreal climates, the gradual disappearance of large extents of permafrost and the reduction of frost periods in extensive belts adjoining former permafrost are expected to improve the internal drainage of soils in vast areas, with probable increases in leaching rates. The appreciable increase in period when the soil temperature is high enough for microbial activity would lead to lower organic matter contents, probably not fully compensated by increased primary production through somewhat higher net photosynthesis and the longer growing period. Paradoxically, the extent of soils subject to periodic reduction could well increase in level areas, in spite of the greater leaching, because of increased periods when the soils are water-saturated but also sufficiently warm for microbial activity. Soils most resilient against such effects, including the leaching of nutrients and periodic soil reduction, would have similar characteristics as the most resilient ones in other climates: adequate cation exchange capacity and anion sorption to minimize nutrient loss during leaching flows, a high structural stability and a strongly heterogeneous system of continuous macropores to maximize rapid bypass flow during periods with excess meltwater.

Processes in Soils

The most rapid processes of chemical or mineralogical change under changing external conditions would be loss of salts and nutrient cations where leaching increases, and salinization where net upward water movement occurs because of increased evapotranspiration or decreased rainfall or irrigation water supply. The clay mineral composition and the mineralogy of the coarser fractions would generally change little, even over centuries. Exceptions would be the transformation of X-ray amorphous material into the clay mineral halloysite when a volcanic soil previously under perennially moist conditions becomes subject to periodic drying, or the gradual dehydration of goethite to haematite in soils subject to higher temperatures or severe drying, or both. Changes in the surface properties of the clay fraction, while generally slower than salt movement, can take place much faster than changes in bulk composition or crystal structure. Such surface changes have a dominant influence on soil physical and chemical properties.

Changes in the clay mineral surfaces or the bulk composition of the clay fraction of soils are brought about by a small number of transformation processes, listed below. Each of these processes can be accelerated or inhibited by changes in external conditions due to global change.

1. hydrolysis by water containing carbon dioxide, which removes silica and basic cations;
2. cheluviation, which dissolves and removes especially aluminium and iron by chelating organic acids;
3. ferrolysis, a cyclic process of clay transformation and dissolution mediated by alternating iron reduction and oxidation, which decreases the cation exchange capacity by aluminium interlayering in swelling clay minerals;
4. dissolution of clay minerals by strong mineral acids, producing acid aluminium salts and amorphous silica;
5. reverse weathering, i.e., clay formation and transformation under neutral to strongly alkaline conditions, which may create, e.g., montmorillonite, palygorskite or analcime.

Hydrolysis and cheluviation may be accelerated by increased leaching rates. Ferrolysis may occur where soils are subject to reduction and leaching in alternation with oxidation: in a warmer world, this may happen over larger areas than at present, especially in high latitudes and in monsoon climates. Dissolution by strong acids would occur, e.g., where sulphidic materials in coastal plains are oxidized with an improvement of drainage; however, a rise in sea level would reduce the likelihood of this occurring naturally. Reverse weathering could begin in areas drying out during global warming, and would continue in most presently arid areas.

These processes would influence the surface properties of the clay fraction only over a period of centuries, even with the changes envisaged as a consequence of global warming. By contrast, direct human action can vastly accelerate some of these processes as is evident, for example, from the severe effects of acid rain on sandy soils in parts of Europe or from the extremely rapid ferrolysis in soils seasonally inundated by water level fluctuations in the Volta lake in Ghana.

Not only the speed of soil formation can be accelerated by human action, but also, albeit much more locally, its very nature or direction. In most places, the natural soil-forming processes are not fundamentally changed, but there are certain threshold situations, generally with fragile soils, where even a small change in external conditions may cause a major, and adverse, change from one dominant soil-forming process to another. The four examples summarized below illustrate a change from hydrolysis to cheluviation (Ferralsols to Podzols); irreversible hardening of the subsoil; clay illuviation forming a dense subsoil in originally homogeneous, porous Ferralsols; and salinization.

The yellowish sandy Ferralsols and Ferralic Arenosols of Eastern Amazonia, Kalimantan and the Zaire basin may rapidly change into Podzols or Albic Arenosols (giant Podzols) with even small increases in total rainfall or stronger seasonality, or increased input of acidic ('poor') organic matter. An increase in effective rainfall due to climate change may cause a major increase in the extent of Podzols formed from present-day yellowish sandy Ferralsols where, presently, Podzols occur in patches within the Ferralsols area.The imperfectly drained loamy Plinthosols on the flat interfluves of Western Amazonia would change into shallow, droughty soils with an irreversibly hardened subsoil if subject to drying out with climate change.

The deep reddish, porous loamy to clayey Ferralsols of the transition zones between forest and savanna in Eastern Africa, stable under the present vegetation, may be leached so far that a denser subsoil with washed-in clay is formed below an unstable topsoil with little organic matter, as already observed where the land was cleared several decades ago; the same may happen over more extensive areas under a sparser vegetation brought about by a somewhat drier climate.

The silty Fluvisols in the broad river valleys of the Sudano-Sahelian zone of West Africa, such as the interior delta of the Niger river, may become saline or sodic upon even minimal change in precipitation and flooding regimes - as exemplified by current human actions with the same soil-hydrological implications.

Some Properties of Clay Surfaces

Soils with a naturally high structural stability, for example Ferralsols and Nitisols, occupy sizeable areas in the tropics. The former are widespread, among others, in northern South America, the latter in East Africa. The clay fraction in such soils generally has oxidic surfaces: mainly iron(III) and aluminium oxides or hydroxides, while the bulk of the clay fraction may have different compositions. The oxidic surfaces could form from parent materials with moderate or high iron contents under long-continued hydrolysis by water (containing carbon dioxide).

At the other extreme are soils with a very low structural stability, or with a severe hazard of failure under load or shock (so-called quick clays). The surfaces of the clay minerals in these soils are generally covered by amorphous, gel-like material with a high silica content. Such material may have originated in earlier periods when the soils were strongly saline and, presumably, subject to processes of reverse weathering. Examples are the quick clays of the Champlain Sea sediments in Ontario and Quebec and of parts of Scandinavia. Such soils are most likely to generate high proportions of runoff and suspended sediment, but also most liable to mudflows once sloping sites are water-saturated to appreciable depth. Some Andosols are thixotropic and have similarly low stability because of their similar composition, derived from volcanic materials (tuff).

Most soils fall somewhere between these extremes. Vertisols, for example, have moderate or low structural stability, and clay surfaces that are mainly silica, but with generally small amounts of amorphous coating. In Planosols, if formed by ferrolysis, the clay fraction in the upper, eluvial horizons has been partly decomposed with a residue of amorphous silica, but the remaining smectite or illite has been interlayered with aluminium hydroxide polymers, which has decreased the swell-shrink potential and the cation exchange capacity of the clay fraction. Concurrently, parts of the free iron oxides have been reduced and leached out. The net effect of these changes generally is a decrease in structural stability.

Resilience against Physical and Chemical Soil Degradation

As discussed, most soils do not have a high intrinsic resilience against physical soil degradation by, for example, high-intensity rainfall. In natural conditions in humid climates, it is the complete soil cover near ground level combined with the perforating activity of the soil fauna that makes the soil-vegetation system resilient against physical degradation.

In the Rhine river plain in the Netherlands, for example, most of the originally calcareous alluvial soils have been decalcified within a millennium or so. Only in small areas on the highest levees of that age that have continually remained under forest, soils are still calcareous, and even have lime pseudomycelia (filaments) indicative of less humid soil conditions, and abundant vertical macropores produced by earthworms. In these soils, faunal activity is high because of the adequate litter supply: the resulting macropores remain open, protected against rain impact by litter and undergrowth: and excess water from heavy rain passes to the substratum through the macropores without leaching lime from most of the soil mass.

Research to increase resilience of soils against any adverse effects of climate change could be done in conjunction with that on soil resilience against direct adverse human impacts. Until site-specific management procedures have been elaborated, soil and crop (including trees and pasture) management should aim to maintain soil cover and organic matter supply to soil biota, while minimizing mechanical disturbance by heavy traffic, cultivation or excessive grazing intensity. Such kind of management may also help to conserve plant nutrients (in soils not flooded for wetland cultivation) since the stable, heterogeneous system of biopores produced by the soil fauna would favour bypass flow of any excess moisture and thus decrease leaching through the soil mass.

A single management recipe would not be generally applicable in different conditions. Minimizing damage by certain termite species harming crop performance may necessitate a period without residues on the soil, for example; or crop residues may be needed for feed or fuel. Management methods for wetland need to be developed that make optimum use of any increased potential productivity, while minimizing secondary effects such as increased CH_4 or N_2O emission from the reduced soil. Such factors, and others, should be taken into account in designing an optimum management strategy for any specific natural and cultural environment.

Resilience Against Soil Reduction

Soil reduction, which would limit land suitability for dryland crops, or strong reduction, which would be liable to produce toxins even for wetland crops, may take place once the soil is water-saturated long enough for microbial action to exhaust the oxygen remaining in the soil when water-saturation started. Another necessary condition is the presence of sufficient readily decomposable organic matter as an energy source for the microbial activity. In most soils, during reduction the redox status is stabilized at an Eh about 100-200 mV near neutrality by the Fe^{2+} - $Fe(OH)_3$ equilibrium, except where the content of readily decomposable organic matter is very high or the content of free iron (III) oxides very low. In such cases, negative Eh values may occur, and toxic hydrogen sulphide or low-molecular organic compounds - including methane - may be formed.

Resilience against soil reduction in practice depends on the drainage conditions, since most soils have sufficient organic matter for reduction to start within about a week after water-saturation. Soils most resilient against reduction in conditions of increased rainfall variability and incidence of high-intensity rainfall have similar properties as those resilient against the negative effects of other perturbations: high infiltration rate, high structural stability and a permanent heterogeneous system of tubular macropores, good external drainage.

Soil Reaction (pH)

Most soils would not be subject to rapid pH changes resulting from climate change. Exceptions might be found in potential acid sulphate soils, extensive in some coastal plains and estuaries, if they become subject to increasingly long dry seasons. Even though most of such soils are clays with moderate or high cation exchange capacity, the amounts of acid liberated in such soils upon oxidation generally exceed this rapid buffering capacity. Therefore, pH values may temporarily reach 2.5 to 3.5 and a small part of the clay fraction may be decomposed as indicated under *Processes in soils,* above. This then buffers the pH generally between 3.5 and 4 in the long run. Depending on the efficiency with which the excess acid formed can be leached out, the period of extreme acidity and aluminium toxicity may last between less than a year and several decades.

In calcareous soils, soil reaction may range between about 8.5 and 7 depending on the partial pressure of CO_2 in the soil; this range is maintained against leaching of basic cations by the different soil processes as long as a few per cent of finely distributed lime remain. Buffering in non-calcareous soils is less strong, but depends on the cation exchange capacity at soil pH. In soils with variable-charge surfaces of the clay fraction, this decreases with acidification.

It should be noted that the simple modelling of accelerated $CaCO_3$ leaching under a doubled atmospheric CO_2 concentration generally does not hold true. In most soils, the ongoing decomposition of organic matter maintains CO_2 concentrations in the soil air far above atmospheric concentration even now, and $CaCO_3$ solubility is determined by the partial pressure of CO_2 in soil air and its activity in soil water, rather than in the atmosphere. Leaching of lime is thus positively related to rate of organic matter decomposition, negatively to gas diffusion rate, and positively to amount of water percolating through the soil.

In conditions where leaching is accelerated by climate change, it would be possible to find relatively rapid soil acidification after a long period with little apparent change, as has been the case - but after a shorter latent period - in some soils in Europe that have been subject to acid rain for several decades. The soil might in fact be steadily depleted of basic cations, but a pH change may start, or may become more rapid, once certain buffering pools are nearly exhausted. Such non-linear and time-delayed effects have been discussed in the context of soil and water pollution by Stigliani; they are also expected to occur in various ways at different times after increased temperatures and changed rainfall patterns will have been operative.

Effects of a Rising Sea Level on Soils in Coastal Areas

The probable effects on soil characteristics of a gradual eustatic rise in sea-level will vary from place to place depending on a number of local and external factors, and interactions between them. In principle, a rising sea level would tend to erode and move back existing coastlines. However, the extent to which this actually happens will depend on the elevation, the resistance of local coastal materials, the degree to which they are defended by sediments provided by river flow or longshore drift, the strength of longshore currents and storm waves, and on human interventions which might prevent or accelerate erosion.

In major deltas, such as those of the Ganges-Brahmaputra and the major Chinese rivers, sediment supplies delivered to the estuary will generally be sufficient to offset the effects of a rising sea level. Such deltaic aggradation could decrease, however, under three circumstances:

1. where human interventions inland, such as large dams or successful soil conservation programmes, drastically reduce sediment supply to the delta: e.g., the construction of the Aswan high dam in 1964 has led to coastal erosion and increased flooding of lagoon margins in the Nile delta;
2. where construction of embankments within the delta interrupts sediment supply to adjoining backswamps, exposing them to submergence by a rise in sea level: e.g., embankments along the lower Mississippi river have cut off sediment supplies to adjoining wetlands which formerly offset land subsidence occurring due to compaction of underlying sediments;
3. where land subsidence occurs due to abstraction of water, natural gas or oil: e.g., as is presently happening in Bangkok and in the northern part of the Netherlands.

In coastal lowlands which are insufficiently defended by sediment supply or embankments, tidal flooding by saline water will tend to penetrate further inland than at present, extending the area of perennially or seasonally saline soils. Where *Rhizophora* mangrove or*Phragmites* vegetation invades the area, that would over several decades lead to the formation of potential acid sulphate soils. Impedance of drainage from the land by a higher sea level and by the correspondingly higher levels of adjoining estuarine rivers and their levees, will also extend the area of perennially or seasonally reduced soils and increase normal inundation depths and durations in river and estuary basins and on levee backslopes. In sites which become perennially wet, soil organic matter contents will tend to increase, resulting eventually in peat formation. On the other hand, where coastal erosion removes an existing barrier of mineral soils or mangrove forest, higher storm surges associated with a rising sea level could allow sea-water to destroy existing coastal eustatic peat swamps, which might eventually be replaced by freshwater or saltwater lagoons.

The probable response of low-lying coastal areas to a rise in sea level can be estimated in more detail on the basis of the geological and historical evidence of changes that occurred during past periods when sea level was rising eustatically or in response to tectonic or isostatic movements: e.g., around the southern North Sea; in the Nile delta; on the coastal plain of the Guyanas; in the Musi delta of Sumatera. Contemporary evidence is available in areas where land levels have subsided as a result of recent abstraction of water, natural gas or oil from sediments underlying coastal lowlands. Further studies of such contemporary and palaeoenvironments are needed together with location-specific studies in order to better understand the change processes, identify appropriate responses and assess their technical, ecological and socio-economic implications

Some major and widespread soil changes expected as a result of any global change are positive, especially the gradual increases in soil fertility and physical qualities consequent on increased atmospheric CO_2 The increased productivity and water-use efficiency of crops and vegetation, and the generally similar or somewhat higher rainfall indicated by several global circulation models, not fully counteracted by higher evapotranspiration, would be expected to lead to widespread increases in ground cover, and consequently better protection against runoff and erosion.

Major but less widespread soil changes, including greater biological activity and increased extent of periodic reduction in soils, would be expected where permafrost would disappear. In unprotected low-lying coastal areas, gradual encroachment of *Rhizophora* mangroves or*Phragmites* following more extensive brackish-water inundation may give rise to the formation of potential acid sulphate soil layers after several decades. Deeper and longer-duration flooding of basins and levee backslopes in adjacent river and estuary plains could lead to more extensive reducing conditions and increased organic matter contents, and locally to peat formation.

Other changes due to climate change (temperature and precipitation) are expected to be relatively well buffered by the mineral composition, the organic matter content or the structural stability of many soils. However, decreases in cover by vegetation or annual or perennial crops, caused by any locally major declines in rainfall not compensated by CO_2 effects, could lead to soil structure degradation and decreased porosity, as well as increased runoff and erosion on sloping sites and by the concomitant more extensive and rapid sedimentation. Changes in options available to land users because of climate change may have similar effects.

In certain fragile soils, the nature of the dominant soil-forming process may change for the worse with increased, decreased or more strongly seasonal rainfall.

In most cases, changes in soils by direct human action, on-site or off-site (whether intentional or unintended), are far greater than the direct climate-induced effects. Soil management measures designed to optimize the soil's sustained productive capacity would therefore be generally adequate to counteract any degradation of agricultural land by climate change. Soils of nature areas, or other land with a low intensity of management such as semi-natural forests used for extraction of wood and other products, are less readily protected against the effects of climate change but such soils, too, are threatened less by climate change than by human actions - off-site, such as pollution by acid deposition, or on-site, such as excessive nutrient extraction under very low-input agriculture.

To armour the world's soils against any negative effect of climate change, or against other extremes in external circumstances such as nutrient depletion or excess (pollution), or drought or high-intensity rains, the best that land users could do, would be:

1. to manage their soils to give them maximum physical resilience through a stable, heterogeneous pore system by maintaining a closed ground cover as much as possible;
2. to use an integrated plant nutrient management system to balance the input and offtake of nutrients over a cropping cycle or over the years, while maintaining soil nutrient levels low enough to minimize losses and high enough to buffer occasional high demands.
3. An analogous philosophy, at lower levels of external inputs, could be formulated for extensive grazing land and production forest, whether planted or managed natural forest.

Human action and management has been emphasized in these conclusions because most of the world's land is used and, to different degrees, managed rather than under natural conditions.

References

Amatekpor, J.K. 1989. The effect of seasonal flooding on the clay mineralogy of a soil series in the Volta lake drawdown area, Ghana. *Land Degradation and Rehabilitation* 1: 89-100.

Brammer, H. and Brinkman, R. 1990. Changes in soil resources in response to a gradually rising sea-level. Chapter 12. In: Scharpenseel *et al.* (eds.). 1990. pp. 145-156.

Emanuel, W.R., Shugart, H.H. and Stevenson, M.P. 1985. Climatic change and the broad-scale distribution of terrestrial ecosystem complexes. *Climatic Change* 7: 29-43.

Warrick, R. and Farmer, G. 1990. The greenhouse effect, climatic change and rising sea level: implications for development. *Trans. Inst. Br. Geogr. N.S.* 15: 5-20.

5

The CO_2 Fertilization Effect

The rise in atmospheric carbon dioxide (CO_2) concentration from about 280 μ mol/mol before the industrial revolution to about 360 μ mol/mol currently is well documented. The consensus of many studies of the effects of elevated CO_2 on plants is that the CO_2 fertilization effect is real. However, the CO_2 fertilization effect may not be manifested under conditions where some other growth factor is severely limiting, such as low temperature. Also, plants grown in some conditions, where limitations of rooting volume, light, or other factors restrict growth, have not shown a sustained response to elevated CO_2.

CO_2 Effects on Plant Growth Processes

Most of the following discussion of CO_2 effects on plants applies to species with the C_3 photosynthetic pathway and not necessarily to species with the C_4 pathway. Other aerial, non-biotic environmental factors that affect plant growth and development are light and temperature. Plant photosynthetic rates generally increase linearly with light across relatively low ranges of light intensity, and then the rates decelerate until they reach an asymptotic maximum. Because of crowding and shading of many leaves, most crop canopies do not reach light saturation at full sunlight; that is, they would be able to respond to light levels well beyond full solar irradiance. Likewise, crop photosynthetic rates respond to increasing levels of CO_2 but then level off at higher concentrations (around 700 μ mol/mol or greater, depending upon species and other factors). However, leaf photosynthesis usually increases with temperature up to some maximum value, and then declines. Furthermore, temperature affects not only photosynthesis, but also respiration, growth, development phases and reproductive processes.

Elevated CO_2 may have some effects on crop phenology, although stages of development are governed primarily by temperature, time and photoperiod. If dates of planting were to be changed because of the greenhouse effect, then phenological timing of plants could be affected. For example, higher temperatures could decrease yields by decreasing the duration of the grain-filling period or changes in photoperiod could shorten or lengthen the vegetative stage.

The CO_2 fertilization effect begins with enhanced photosynthetic CO_2 fixation. Non-structural carbohydrates tend to accumulate in leaves and other plant organs as starch, soluble carbohydrates or polyfructosans, depending on species. In some cases, there may be feedback inhibition of photosynthesis associated with accumulation of non-structural carbohydrates. Increased carbohydrate accumulation, especially in leaves, may be evidence that crop plants grown under CO_2 enrichment may not be fully adapted to take complete advantage of elevated CO_2. This may be because the CO_2-enriched plants do not have an adequate sink (inadequate growth capacity), or lack capacity to load phloem and translocate soluble carbohydrates. Improvement of photoassimilate utilization should be one goal of designing cultivars for the future.

In the process of growth, photoassimilates are allocated to the vegetative shoots, root system or reproductive organs. In some cases, more photoassimilate of CO_2-enriched plants is partitioned to the root system than to the shoots. Above ground, more photoassimilate usually goes into stems and supporting structures than into leaves. This phenomenon may not be an inherent response to elevated CO_2, but may be a by-product of the larger size of plants often found in CO_2-enriched atmospheres, especially by species that produce branch stems along the aerial mainstems.

Reproductive biomass growth as well as vegetative biomass growth are usually increased by elevated CO_2. However, the harvest index, or the ratio of seed yield to above-ground biomass yield, is typically lower under elevated CO_2 conditions, which may also be evidence of the lack of capacity to utilize completely the more abundant photoassimilate.

In many cases, both the amount and the carboxylation activity of ribulose 1,5-bisphosphate carboxylase-oxygenase enzyme (rubisco) is decreased in leaves of plants grown under elevated CO_2 This acclimation phenomenon may produce 'downregulation of photosynthesis'; however, this is not universally the case. For example, there is little evidence of this downregulation response in soybean (*Glycine max* L. Merr.), a C_3 legume. In fact, photosynthetic capacity per unit leaf area of soybean is increased under CO_2 enrichment. Leaves often develop an additional layer of mesophyll cells. Also, more structural carbohydrate may be produced in leaves, as well as stems, of CO_2-enriched plants.

We can obtain clues about reasons that some plants downregulate and others do not in response to elevated CO_2 by focusing on the capabilities of soybean. This plant has (a) symbiotic N2 fixation; (b) the capacity to form additional layers of palisade cells in the leaf tissue; (c) the capacity to shunt much of the photoassimilate into relatively inert starch rather than soluble sugars during photosynthesis; (d) a relatively strong leaf and stem sink during vegetative development; and (e) a strong seed-fill sink during reproductive development. Plants which lack these capacities, either inherently or because of growth in limiting environments, are more likely to demonstrate some degree of downregulation of photosynthesis.

The carbon:nitrogen ratio of leaves of plants is usually increased under CO_2 enrichment. Plants may acclimate to elevated CO_2 by requiring less rubisco and photo-synthetic apparatus, which would lead to lower nitrogen contents. The overall change in C:N ratios is governed both by increases in structural and non-structural carbohydrates, and by decreases in protein content. However, seed nitrogen content is little affected.

Specific respiration rates may be reduced by both short-term exposure to elevated CO_2 and long-term growth at elevated CO_2. However, the long-term effect may be similar when respiration rates are reported on a per unit nitrogen basis.

In climate change scenarios, temperatures are predicted to increase following the rise of CO_2 and other greenhouse-effect gases. Carbon dioxide x temperature interactions have been observed for vegetative growth (i.e., the CO_2 fertilization effect is greater at warmer temperatures than at cooler temperatures). Temperature increases in a higher CO_2 world could increase overall biomass productivity for vegetative crops (pastures and forages) both by extending the length of the growing season in temperate regions, and by the interaction of CO_2 x temperature in stimulation of vegetative growth.

However, CO_2 x temperature interactions appear to be very small or negligible for reproductive processes (seed set and seed yield) although there may be more initial flowers formed by greater amounts of branching or tillering that is stimulated by CO_2 enrichment.

Precipitation changes may occur along with other climatic change effects. In general, predictions from crop models show that increased CO_2 should increase productivity of C_3 plants, but the associated predictions of temperature rise will be detrimental. Not surprisingly, changes in precipitation patterns (decreases of rainfall during growth period) could be more detrimental for crop production than changes in temperature.

Under elevated CO_2 stomatal conductance in most species will decrease which may result in less transpiration per unit leaf area. However, leaf area index of some crops may also increase. The typical 40% reduction in stomatal conductance induced by a doubling of CO_2 has generally resulted in only a 10% (or less) reduction in crop canopy water use in chamber or field experimental conditions. Actual changes in crop evapotranspiration will be governed by the crop energy balance, as mitigated by stomatal conductance, leaf area index, crop structure and any changing meteorological factors.

Water-use efficiency (WUE) (ratio of CO_2 uptake to evapotranspiration) will increase under higher CO_2 conditions. This increase is caused more by increased photosynthesis than it is by a reduction of water loss through partially closed stomata. Thus, more biomass can be produced per unit of water used, although a crop would still require almost as much water from sowing to final harvest. If temperatures rise, however, the increased WUE caused by the CO_2 fertilization effect could be diminished or negated, unless planting dates can be changed to more favourable seasons.

Several assessments of impacts of climate change on crop productivity have been published. Progress has been made on integrating the impacts on individual countries and on economic and social interactions. For the most part, these assessments project more favourable climates for agriculture in northern latitudes and less favourable climates in the tropical and subtropical zones. However, the crop modelling predictions are dependent on the scenarios of outputs of General Circulation Models (GCMs) applied to the greenhouse effect. Thus, the dependency chain of assessments follows: Climate Change Scenarios $\rightarrow$Crop Model Prediction and Agricultural

Production Systems (with and without available mitigation and adaptation response strategies) →National Scenarios of Economics and Well-being of Farmers, Agricultural Commerce, and Consumers →Country-by-Country and Global Interaction Scenarios of World Trade (food and all other commodities). Population Dynamics and Economic Well-being, and Impacts on Social Systems.

As the world continues to consume fossil fuels, CO_2 concentrations will continue to rise. Other greenhouse-effect gases, such as methane, nitrous oxides, chlorofluorocarbons and chlorofluorocarbon substitutes, and perhaps tropospheric ozone, will likely rise also. The CO_2 fertilization effect on plants will increase and climate changes may occur because of the combined increase of all greenhouse-effect gases. Global agriculture could adapt to gradual regional climate changes, but sudden changes would be more serious. Adaptation and/or mitigation actions could include the following:

1. Selection of plants that can better utilize carbohydrates which are produced when plants are grown at elevated CO_2.
2. Selection of plants that produce less structural matter and more reproductive capacity under CO_2 enrichment. (This applies for seed crop plants, not necessarily vegetative biomass plants.)
3. Search for germplasms that are adapted to higher day and night temperatures, and incorporate those traits into desirable crop production cultivars to improve flowering and seed set.
4. Change planting dates and other crop management procedures to optimize yields under new climatic conditions, and select for cultivars that are adapted to these changed agricultural practices.
5. Shift to species that have more stable production under high temperatures or drought.
6. Determine whether more favourable N:C ratios can be attained in forage cultivars adapted to elevated CO_2.
7. Where needed, and where possible, develop irrigation systems for crops.

Specific Responses of Crops to Elevated CO_2

This section will focus on responses of two C_3 crops to elevated CO_2: soybean, a symbiotic nitrogen-fixing legume representative of the pulses; and rice (*Oryza sativa* L.), a globally important cereal food crop. Findings from seven crop cycles of both soybean and rice grown in sunlit chambers at Gainesville, Florida, USA, will be emphasized. The chambers feature both large rooting volumes of real soil and light from the sun. Some studies included responses to sub-ambient as well as superambient CO_2

Soybean

Photosynthetic rates

Soybean photosynthetic rates of both leaves and the whole canopy have *always* been increased by elevated CO_2 Jones *et al.* reported midday maximum canopy photosynthetic rates of 60 and 90 μ mol $CO_2/m^2/s$ at 70 days after planting (DAP) for soybean grown at CO_2 concentrations of 330 and 800 μ mol/mol, respectively. The leaf area index (LAI) was 6.9 and 9.0, respectively. Midday maximum canopy photo-synthetic rates in other studies were 40 and 75 μ mol/m^2/s for soybean grown at 320 and 640 μ mol/mol CO_2 and 40 and 80 μ mol/m^2/s for soybean grown at 330 and 800 μmol/m^2/s.

Valle *et al.* found that midday maximum photosynthetic CO_2 uptake rates of soybean leaves ranged from 30 to 50 μ mol/m^2/s and 15 to 25 μ mol/m^2/s on plants grown at 660 and 330 m mol/mol CO_2 respectively. Allen *et al.* reported that, at all light levels, leaf photosynthetic rates increased linearly with CO_2 concentration across the range of 330 to 800 μ mol/mol.

Valle *et al.* used a Michaelis-Menten type of rectangular hyperbola to summarize photosynthetic responses of soybean leaves vs. CO_2 concentration. The plants had been grown at 330 and 660 μ mol/mol of CO_2 and then exposed to a wide range of CO_2 for a short period.

$$Y=(Y_{max} \text{ x } [C])/([C]+K_m)+Y_i \quad (1)$$

where Y is photosynthetic rate in μ mol $CO_2/m^2/s$; $[C_3]$ is CO_2 concentration in μ mol/mol; Y_i is the y-axis intercept at zero $[C_3]$, the apparent respiration rate, in μ mol $CO_2/m^2/s$; Y_{max} is the response limit of $(Y - Y_i)$ at very high [C], the asymptotic photosynthetic rate, in mmol $CO_2/m^2/s$; K_m is the value of [C] where $(Y-Y_i)=Y_{max}/2$, the apparent Michaelis-Menten constant, in (μ mol/mol; and G_c is the calculated [C] intercept at zero Y, the CO_2 compensation point, μ mol/mol (not shown in this equation). There was no obvious downregulation of soybean leaf photosynthesis in response to elevated CO_2; in fact, photosynthetic capacity was increased. Leaf quantum yield increased from 0.05 to 0.09 in the soybean leaves exposed to CO_2 of 330 and 660 μ mol/mol, respectively.

Campbell *et al.* showed that soybean leaf photosynthetic rates were higher for plants grown at 660 than at 330 μ mol/mol CO_2 when measured at common intercellular CO_2 concentrations. Furthermore, Campbell *et al.* measured rubisco activity and amount in leaves of soybean grown in CO_2 concentrations of 160, 220, 280, 330, 660 and 990 μ mol/mol. They found that rubisco activity was almost constant at 1.0 μ mol CO_2/min/mg soluble protein across this CO_2 treatment range. Leaf soluble protein was nearly constant at about 2.4 g/m^2 with 55% being rubisco protein. Specific leaf weight increased across the 160 to 990 μ mol/mol CO_2 concentration range, so that the rubisco activity on a leaf dry weight basis decreased.

Campbell *et al.* showed that photosynthetic capacity of soybean canopies grown at 330 and 660 μ mol/mol CO_2 were similar for short exposures at concentrations below 500 m mol/mol, but above this concentration, canopies grown at 660 μ mol/mol had a slightly higher photosynthetic capacity. Thus, soybean did not lose photo-synthetic capacity as did some other plant species. In other studies, however, both photosynthetic rate and rubisco activity of soybean declined during long-term CO_2 enrichment.

Respiration

Whole-canopy pre-dawn respiration rates at 40 to 60 DAP were 2-3 and 4-5 μ mol/m^2/s for soybean grown at CO_2 levels of 320 and 640 μ mol/m^2, respectively, and a constant day/night temperature of 25°C. After cross-switching two of the treatment chambers at DAP 52, the subsequent canopy photosynthetic rates and respiration rates quickly adjusted to their new CO_2 exposure conditions.

Partitioning

Growth of plants under elevated CO_2 results in changes in partitioning of photoassimilates to various plant organs over time. In soybean, elevated CO_2 generally promoted greater carbon (dry matter) partitioning to the supporting structure (stems, petioles and roots) than to the leaf laminae during vegetative stages of growth. During reproductive stages, there tended to be lower relative partitioning to reproductive growth (pods) by plants under elevated CO_2.

Growth rates

During the linear phase of vegetative growth after full ground cover is reached, the growth rates of plants exposed to a range of CO_2concentrations varied from 5.0 to 20.7 $g/m^2/d$ for exposures from 160 to 990 μ mol/mol. The total final dry weight ranged from 12.88 to 39.12 g/plant, and final seed weight ranged from 5.77 to 17.85 g/plant for CO_2 treatments ranging from 160 to 660 μ mol/mol.

Carbohydrates

Soybean accumulates non-structural carbohydrates, particularly starch, under CO_2 enrichment. Allen *et al.* grew soybean under CO_2treatments of 330, 450, 600 and 800 μ mol/mol. Sucrose, reducing sugars, and total soluble sugars of leaves remained somewhat constant throughout the day, but starch increased steadily at a rate of about 6 g/kg dry matter/hour. Average total soluble sugars increased from 24 to 36 g/kg dry weight and starch increased from 85 to 204 g/kg dry weight across the range of 330 to 800 μ mol/mol. Elevated non-structural carbohydrates in CO_2-enriched soybean plants were confirmed by Baker *et al.* and Allen *et al.* The concentrations also varied across the life cycle of the plants.

Nitrogen

Nitrogen content decreased from 50 to 37 g/kg dry weight over the range of 330 to 800 μ mol/mol. When the nitrogen content was adjusted to remove the effect of total non-structural carbohydrate, the relative changes were smaller, 55 to 48 g N/kg dry matter, across the 330 to 800 μ mol/mol range.

Yield

Soybean seed yield was always increased by elevated CO_2 Allen et al. summarized the photosynthetic, biomass and seed yield responses of several experiments with the equation (1) rectangular hyperbola model using data normalized to responses obtained at 330μ mol/mol. The values of K_m, Y_{max} and Y_i parameters for relative photosynthetic rates were 279 μ mol/mol, 3.08 and -0.68, respectively, for relative biomass yield were 182 μ mol/mol, 3.02 and -0.91, respectively, and for relative seed yield were 141 μ mol/mol, 2.55 and -0.76, respectively. This model was used to project yields across several ranges of atmospheric CO_2 concentration increases. For a doubling of CO_2 this model predicted a 32.2% increase in soybean grain yield and a 42.7% increase in biomass. The ratio of these two numbers,

1.322/1.427 = 0.926, gives the fraction of the harvest index expected under doubled CO_2 in comparison with ambient CO_2.

Carbon dioxide and temperature

The CO_2 fertilization effect appears to be enhanced under elevated temperatures, at least up to a point. Idso *et al.* and Kimball *et al.* showed that the growth modification factor (or biomass growth modification ratio) due to a 300 μ mol/mol enrichment was 0.08 per °C (average daily temperature) across the range of 12 to 34°C. However, for soybean, Allen calculated a biomass growth modification ratio response to temperature of -0.031 per °C for seed biomass yield and -0.026 per °C for total biomass accumulation. Elevated temperatures tended to shorten the grain-filling period of this crop.

Soybean seed yield tended to decrease slightly with temperature over the day/night range of 26/19 to 36/29°C. The number of seed per plant increased slightly with increase of both CO_2 and temperature. Mass per seed decreased sharply with increasing temperature. Although CO_2 enrichment resulted in increased seed yield and above-ground biomass, harvest index was decreased with both CO_2 and temperature.

Subsequent experiments with soybean substantiate these data although it was found that seed yields dropped sharply for day/night temperatures of 40/30°C and above, and biomass yields were maintained up to 44/34°C before rapidly failing.

The sensitivity of the growth modification factor data of Idso *et al.* and Kimball *et al.* may have been affected by other factors at Phoenix, Arizona. Apparently, the growth vs. temperature data were obtained throughout the year, and the findings may have been impacted by factors such as total solar radiation, photoperiod, stages of development, or other conditions caused by the changing seasons. Of course, these other environmental factors are all part of the complex of plant responses to climatic conditions. However, a better test of temperature effects alone would be CO_2 enrichment throughout the season or life cycle of plants under natural conditions, but with consistent temperature differences (cooler and warmer).

Evapotranspiration and water-use efficiency

The direct effect of increasing temperatures across the range of 28 to 35°C appears to increase transpiration rate about 4 to 5% per °C, based on both experimental and modelling studies. This is in close agreement with the rise in saturation vapour pressure of about 6% per °C. Allen et al. showed that the increase in WUE of soybean contributed by increased photosynthesis under elevated CO_2 was much greater than the increase in WUE contributed by decreased transpiration. The fractional increase in WUE attributable to increased photosynthesis and decreased transpiration were about 0.8 and 0.2, respectively

Rice

Leaf and canopy photosynthesis

Leaf photosynthetic rates, in high light, of cv. IR72 rice, determined in 1992 (RICE VI experiment) for plants grown at three day/night temperature regimes, 32/23, 35/26 and 38/29°C (but measured at ambient temperature), averaged 18.8 and 30.4 μ mol $CO_2/m^2/s$ for 330 and 660 μ mol/mol CO_2 treatments, respectively.

Following canopy closure, the rice canopy net photosynthetic rate (Pn) vs. photo-synthetic photon flux density (PPFD) responses were linear and did not approach light saturation in any of the experiments, probably because of the erect leaf orientation of cv. IR30 and high plant populations. Canopy Pn vs. PPFD at 60 DAP for the six CO_2 treatments of the RICE II experiment gave values of about 34, 50, 60, 80, 85 and 90 μ mol $CO_2/m^2/s$ at high light (1600 μ mol photons/m^2/s) for treatments of 160, 250, 330, 500, 660 and 900 μ mol/mol.

Linear canopy Pn responses to PPFD on day 60 of the RICE IV experiment were similar among all temperature treatments (25/18, 28/21, 31/24, 34/27 and 37/30°C) that were exposed to 660 μ mol/mol CO_2 responses of the 330 μ mol/mol treatment were about 25% less. Other studies of canopy Pn have shown only small differences across wide ranges of temperature; e.g., cotton and soybean. There may be two reasons for the lack of a clear canopy photosynthetic response to air temperature across the 25 to 37°C range. Evaporative cooling may lower foliage temperature below air temperature increasingly with increasing air temperature and increasing vapour pressure deficit. The summed response of the photosynthetic rates of leaves to temperature at all exposures of light may broaden the temperature response for the whole canopy photo-synthetic rates.

Baker and Allen fitted whole-day canopy Pn data to a rectangular hyperbola of the form of equation (4.1) for DAP 61 of the RICE II experiment. The parameters were 70.8 μ mol/mol, 3.96 and -2.21, for K_m, Y_{max} and Y_i, respectively, with a relative Pn response ceiling of 1.75 at infinite [C]. The calculated percentage increase in response of whole-day canopy Pn at 660 vs. 330 μ mol/mol was 36%.

Photosynthetic acclimation to CO_2 at the canopy level

To test for acclimation of canopy photosynthetic capacity, Baker *et al.* used rice that was grown at 160, 250, 330, 500, 660 and 900 μ mol/mol CO_2. For half-day periods (mornings) on 62,63 and 64 DAP, common CO_2 setpoints of 160, 330 or 660 μ mol/mol were imposed on each chamber on those respective days. Within each of these short-term CO_2 exposure comparisons, Baker *et al.* and Baker and Allen showed that the short-term canopy net photosynthetic rate decreased with increasing long-term CO_2treatment. The relative effects of the short-term CO_2 exposure were greatest for the lowest short-term CO_2 concentration. For example, when compared at a common short-term CO_2 exposure of 160 μ mol/mol, the canopy photosynthetic rate of the chamber containing the 900 μ mol/mol long-term CO_2 growth treatment was only about one-third of that of the 160 μ mol/mol long-term CO_2 growth treatment. However, a large part of this apparent acclimation effect may be attributed to the greater respiration rates of the plants that had been grown under elevated CO_2.

Rubisco protein percentage was used as evidence of leaf acclimation to a wide range of CO_2 concentrations. For rice, rubisco activity expressed on a leaf area basis decreased by 66% across the 160 to 900 μ mol/mol long-term CO_2 treatments. A major cause of this decline in rubisco activity was a 32% decrease in the amount of rubisco protein relative to other soluble protein.

Although Rowland-Bamford *et al.* reported decreases in both amount and activity of rubisco with increasing CO_2 in rice plants grown across the range of 160 to 900 μ mol/mol, the leaf A/C. response curves gave no indication of a downward acclimation of photosynthetic capacity across the smaller range of 330 to 660 μ mol/mol CO_2. One possibility would be that rubisco is not as limiting for photosynthesis under CO_2 enrichment as would be expected. More work is needed, under a range of CO_2 treatments, to explore the interaction effects of sink capacity, nitrogen nutrition, and other internal CO_2-fixation processes on photosynthetic behaviour and crop yield.

Development stages

In rice, two distinctly different rates of leaf appearance for vegetative and reproductive phases of growth were found as has been observed many times before. Leaf appearance rates (leaves per day) are about twice as great in the vegetative phase as in the reproductive phase.

In the 'late' planted (23 June 1987) RICE II experiment, the number of mainstem leaves at panicle initiation and the final number of mainstem leaves decreased across CO_2 treatments of 160 to 500 μ mol/mol and remained similar from 500 to 900 μ mol/mol. Panicle initiation and boot stage occurred about 12 days earlier in the superambient CO_2 treatments compared with the 160 μ mol/mol treatment. Therefore, plant developmental rate was clearly accelerated with increasing CO_2 up to about 500 mmol/mol.

Temperature effect data were assembled from treatments ranging from 25/18/21 to 40/33/37°C (day/night/paddy water temperatures) in RICE I to RICE V experiments. Air temperature greatly influenced leaf appearance rate, developmental rate, and total growth duration, whereas, across the limited range of 330 to 660 μ mol/mol the effects were comparatively small. No consistent differences in phyllochron interval (days per leaf appearance) between 330 and 660 μ mol/mol were observed, whereas phyllochron interval increased with increasing temperature across the range of 25/18/21 to 40/33/37°C, especially for the reproductive phase. In RICE III, IV and V experiments, anthesis ranged from 0 to 6 days earlier in the 660 vs. 330 μ mol/mol treatments. Also, the number of days to anthesis was shortened by approximately 10 days across the temperature treatment range from 25/18/21 to 34/27/31°C for the RICE IV experiment.

Growth and yield

Across a relatively wide temperature range from 25/18/21 to 37/30/34°C, Baker *et al.* found a broad temperature optimum for biomass production in the mid-temperature ranges. Plants grown at 40/33/37°C were near the upper temperature limit for survival. High temperature spikelet sterility of rice is induced almost exclusively on the day of anthesis when temperatures greater than 35 °C for more than one hour induce a high percentage of sterility. In the 40/33/37°C treatments, plants in the 330 μ mol/mol CO_2 chamber died during internode elongation whereas plants in the 660 μ mol/mol chamber produced small, abnormally shaped panicles that were sterile. Therefore, elevated CO_2 may slightly increase the maximum temperature at which rice plants can survive.

At both CO_2 levels, grain yield was highest in the 28/21/25°C treatment followed by a decline to zero yield in the 40/33/37°C treatment. The CO_2 enrichment from 330 to 660 μ mol/mol increased yield by increasing the number of panicles per plant, whereas the number of filled grains per panicle and individual seed mass were less affected. Temperature effects on yield and yield components were highly significant. The number of panicles per plant increased while the number of filled grains per panicle decreased sharply with increasing temperature treatment. Individual seed mass was stable at moderate temperatures but tended to decline at temperature treatments above 34/27/31°C. Final above-ground biomass and harvest index were increased by CO_2 enrichment while harvest index declined sharply with increasing temperature. Notably, there were no significant CO_2 x temperature interaction effects on yield, yield components, or final above-ground biomass.

Canopy dark respiration rates

Plant respiration rates may be decreased by both short- and long-term exposure to high CO_2 concentrations; however Baker *et al.* found that nighttime canopy dark respiration rates [R_d, μ mol $(CO_2)/m^2$ (ground area)/ s] increased for rice exposed to daytime CO_2 ranging from 160 to 900 μ mol/mol. Similar to photosynthetic rates, R_d increased with CO_2 exposure from 160 to 500 μ mol/mol but levelled off somewhat across the 500 to 900 μ mol/mol range. The R_d of the ambient and superambient CO_2 treatments reached a broad maximum around 30 to 50 DAP whereas the broad maximum of the subambient treatments occurred later around 50 to 70 DAP. The maximum values of R_d were about 6, 8, and 9 μ mol/m^2/s for the 160, 250 and 330 μ mol/mol CO_2 treatments and about 11 to 12m mol/ m^2/s for the three superambient CO_2 treatments.

Specific respiration rate [R_{dw}, μ mol (CO_2)/s/kg (total above-ground dry matter)] decreased exponentially with DAP at all CO_2 exposures and was higher in the subambient (160 and 250 μ mol/mol) than in the ambient (330 μ mol/mol) and superambient (500, 660 and 900 mmol/mol treatments). At each CO_2 exposure level, the patterns of R_{dw} with DAP were very similar to the patterns of plant tissue nitrogen concentration with DAP. Furthermore, R_{dw} was linearly related to total above-ground plant tissue nitrogen concentration [mg (N)/g (DW)] across the range of CO_2 exposures and the six datcs of plant sampling (R_{dw} = -13.0 + 0.952 * [N]; r2=0.9, P=0.01). Baker *et al.* concluded that the CO_2 treatments affected R_{dw} mainly by altering the protein content of the plant tissue. Another explanation is that elevated CO_2 increased the amount of structural and non-structural carbohydrates in the plant tissues, so that a larger proportion of dry matter was sequestered in non-protein materials. This study showed no respiration acclimation to long-term CO_2enrichment (indirect acclimation) of rice that could not be explained by nitrogen concentration of the plant tissues.

Evapotranspiration and water-use efficiency

Both plant transpiration and direct evaporation from the floodwater surface contribute to water use (evapotranspiration, ET) of rice growing in the SPAR chambers. Diurnal trends of ET followed diurnal patterns of solar irradiance. After canopy closure, maximum ET rates and total daytime water losses were about 35 and 30% greater, respectively, from rice grown at 160 μ mol/mol compared to rice grown at 900 m mol/mol CO_2. The ET rates were similar when all chambers were exposed for one-half day to the same CO_2 concentration, which demonstrates the effect of the exposure level of CO_2 on stomatal control of transpiration.

Temperature also has a large effect on diurnal ET rates and daytime water use, mediated primarily through vapour pressure deficit of the air. Solar irradiance also has a large effect, probably through both the energy inputs to the canopy and through the stomatal opening response to light (directly or indirectly). Midday maximum ET rates were about 75 and 35% higher for rice grown at 40/33/37°C and 34/27/31°C, respectively, compared to rice grown at 28/21/25°C.

Water-use efficiency: effects of CO_2

Stomatal conductance decreases with increasing CO_2 concentration which can cause a reduction of both leaf and whole canopy transpiration. However, CO_2 enrichment may also increase canopy leaf surface area for transpiration, thereby offsetting some of the water savings. The leaf area index of the RICE II experiment ranged from 7.6 to 10.8 across the CO_2 treatments from 160 to 900 μ mol/mol. Calculations of WUE increased with increasing CO_2 due to the decline in ET across this CO_2 range and the increase in Pn with CO_2 up to the 500 μ mol/mol treatment.

Water-use efficiency: effects of temperature

In the 28°C air temperature treatment, CO_2 enrichment from 330 to 660 μ mol/mol resulted in 30% increase in daytime total CO_2 uptake and 22%

decrease in ET with a concomitant increase of WUE from 1.42 to 2.37 μ mol (CO_2)/mmol (H_2O), an increase of 67%. Daytime ET was more than doubled from the 25 to the 37°C daytime temperature treatment. WUE declined from the 28 to the 37°C daytime treatment due to the sharp increase in ET and the relatively stable Pn across this temperature range. If WUE were based on grain yield, it would decrease drastically with increasing temperature not only because of increasing ET, but also because of sharply decreasing seed production.

Rising atmospheric CO_2 is likely to benefit rice production by increasing photosynthesis, growth and grain yield while reducing water use and increasing WUE. In warm areas of the world, however, possible future global warming may result in substantial yield decreases because of the sensitivity of flowering and seed set to high temperatures and the possibility of water shortages that may result from increased evapotranspiration.

Several recent symposia proceedings and reviews leave little doubt that crop plants can respond well to elevated CO_2. Poorter compiled information from 156 plant species and found that doubling CO_2provided an average growth increase of 37%.

Poorter imposed two restrictions on his compilations that may have led to larger than expected responses to elevated CO_2. Firstly, plants grown in competition were not included. Secondly, only vegetative stages of plants were compared since compiled data were selected prior to flowering.

A number of studies have shown that vegetative growth responses may be greater than reproductive (seed yield) responses. Therefore, the compiled data of Poorter may give an impression of greater response than would be observed throughout the life cycle. Secondly, crops in field conditions usually are grown in dense populations where they compete for space and light. Under more realistic field conditions, crop plants are likely to respond as a community rather than individual plants, wherein light (solar radiation) becomes a limiting factor for growth. Under these conditions, elevated CO_2 cannot promote horizontal expansion and greater light capture. Although the actual field responses may be less, the CO_2 fertilization effect is clearly well-established.

The CO_2 fertilization effect for forest species has also become firmly established. Wullschleger *et al.* estimated the biotic growth factor for 58 controller-exposure studies of forest tree species which included 398 observations.

Kimball *et al.* discussed the data of Idso *et al.* which related the growth modification factor (or biomass growth modification ratio) caused by a 300 μ mol/mol increase in CO_2 concentration above ambient.

Crop Modelling: Predictions for the Future

Modelling Crop Responses ro CO_2 and Climate Changes

Peart *et al.* and Curry *et al.* predicted growth and yield responses of soybean and maize to doubled-CO_2 climate change scenarios of the southeastern USA. Their simulations used 30 years of baseline weather data (1951-1980) from 19 sites in 11 states. Predicted climate changes of the Goddard Institute for Space Studies (GISS) model and the Geophysical Fluid Dynamics Laboratory (GFDL) model were used to change temperatures, precipitation and solar radiation, month-by-month, for the 30 years of baseline data at each site. These modified baseline weather data provided GISS and GFDL climatic change scenarios. At each site, prevailing cropped soil types, planting dates, and cultivars were used in the simulations.

Baseline, GISS and GFDL scenario temperature and rainfall data that were used by Peart *et al.* were condensed for two representative locations; Columbia, South Carolina, and Memphis, Tennessee). Crop yield responses to climate change were simulated under four conditions: with or without direct CO_2 fertilization effects, and under rainfed or optimum irrigation culture. The Penman-Monteith equation, which contains a term for canopy conductance, was used to compute the effects of elevated CO_2 on canopy transpiration. Simulations were run for doubled-CO_2 conditions with a crop photosynthetic enhancement factor of 1.35 for soybean (a C_3plant) and 1.10 for maize (a C_4 plant). No simulations were run with CO_2 fertilization effects only (without climate change effects).

Simulated soybean seed yields were averaged over 30 years and 19 sites. Under rainfed conditions with climate change effects only, simulated soybean yields under the GFDL scenario were reduced 71 % compared to the baseline climate, whereas the average yields were reduced only 23% under the GISS scenario. Yields under the GFDL scenario were severely impacted because of the rainfall reductions predicted by this model.

Under optimum irrigation conditions, average yields under both the GISS and GFDL scenarios were reduced 18 to 19%. However, in spite of higher temperatures, the irrigated yields under the GISS and GFDL scenarios were about 25% greater than the baseline climate scenario without irrigation.

When CO_2 fertilization effects were included with climate change effects, simulated average yields of the GISS climate scenario were increased 11% under rainfed conditions, whereas yields under the GFDL climate scenario were still decreased (-52%). Under optimum irrigation with CO_2 fertilization effects, yields under both GISS and GFDL scenarios were increased 13 to 14%.

Yields of maize were simulated for doubled-CO_2 fertilization effects at only four weather station locations (Charlotte, North Carolina; Macon, Georgia; Memphis, Tennessee; Meridian, Mississippi). For climatic change effects alone, predicted maize yields declined only 6% in the GISS scenario, but declined 73% in the GFDL scenario. Although irrigation increased predicted crop yields, the GISS and GFDL climate scenarios gave yield decreases of 18 and 27%, respectively, with respect to the irrigated baseline weather. The yield reduction of the GFDL scenario with respect to the GISS scenario was 10%, attributable to slightly higher temperatures. Including the CO_2fertilization effects with climatic change scenarios had little effect on the predicted yields of maize because it is a C_4 plant.

In the Great Lakes and Corn Belt Area, simulations by Ritchie *et al.* showed that higher temperature would have the greatest effect of the climate change factors on predicted yields of soybean and maize, mainly through decreases in the crop life cycle. For the southern part of this region, yield reductions were greater for the GFDL scenario. Predicted yields increased for the northernmost stations because temperatures and growing season duration became more favourable. Average irrigation water requirements in this region increased about 90% for the GISS and GFDL scenarios.

Rosenzweig modelled maize and wheat yields in the Great Plains under GISS and GFDL scenarios. Yields decreased in the Southern Great Plains because higher temperatures shortened the life cycle of the crops. Where precipitation was predicted to decrease, irrigation requirements increased. In a separate modelling study, Allen and Gichuki predicted a 15% increased requirement for irrigation for this region, with greater requirements for alfalfa because its growing season was increased, and lower requirements for maize and winter wheat because their growing seasons were decreased. The CO_2 fertilization effect offset the adverse effects of climate change at some locations of Ritchie *et al.* and Rosenzweig.

Dudek predicted productivity changes of several Californian vegetable, fruit and nut crops in response to GISS and GFDL scenarios for doubled

CO_2. Without CO_2 fertilization effects, statewide average yield changes, depending on the crop, would be -8 to -34% for the GISS scenario and -6 to -31 % for the GFDL scenario. With CO_2 fertilization effects, predicted statewide yield changes ranged from about +17 to -12% for the GISS scenario and about +21 to -8% for the GFDL scenario.

Dudek used a California Agriculture and Resource Model (CARM) to further predict economic and market impact of the productivity changes. Production declined, in general, under the scenarios of climate change without CO_2 fertilization effects; however, commodity prices generally increased. Under the CO_2 fertilization plus climate change effects scenarios, predicted impacts on commodity prices were much less.

The effect of temperature on the phenology of crop plants plays a critical role. One crucial need is for more detailed research on responses of plants to temperature, and temperature x CO_2 interactions, as inputs to crop models. More modelling studies are also needed on different planting dates as an adaptive strategy. Cultivars need to be designed for future climatic conditions. Factors that should be considered are: extension of the grain-filling period and perhaps shortening the duration of vegetative growth (which would also improve harvest index); ability to flower and set seed at higher temperatures; photoperiod and thermoperiod adaptive interactions; selection for positive photosynthetic acclimation where negative photosynthetic acclimation has been observed; and capability of utilizing photoassimilates (storage carbohydrates) more effectively. These factors need to be integrated into whole-plant physiology under real-world conditions.

Adaptations and Evapotranspiration Requirements

The simulated crop yield responses to climatic changes provided by Peart *et al.* and Curry *et al.* manifest two main points: (a) the serious adverse impact of inadequate rainfall scenarios on crop production coupled with rising temperature scenarios, and (b) the importance of beneficial CO_2 fertilization effects in the face of elevated temperatures. However, the climate change for an effective doubling of CO_2 may occur at CO_2 concentrations less than those used in this simulation, if radiatively active trace gases other than CO_2play a large role in the greenhouse effect.

If global warming lags the increases of atmospheric CO_2, then some beneficial effects of CO_2 fertilization are likely to occur before the full impact of climate change is manifested. However, Broecker and others caution that climate changes have not always been gradual during interglacial periods of the Pleistocene Era. There is clear evidence of relatively rapid climatic oscillations in the northern hemisphere during the previous interglacial period 110000 to 140000 years before present based on Greenland Ice-core Project (GRIP) records. These oscillations produced cold periods that were as severe as the proceeding glacial period.

Much of the reduction in soybean yields reported by Peart *et al.* and Curry *et al.* was caused by decreases in the length of the grain-filling period under higher temperatures. Changes in management practices, such as changing planting dates or selection of other cultivars, may help to prevent some of the potential reductions in yield. In the future, plant breeders may need to adapt combinations of temperature tolerance and photoperiod responses into new germplasm. In situations where non-structural carbohydrates accumulate as a CO_2 fertilization effect response, new germplasm needs to be developed that can make better use of the photoassimilate source.

Irrigation is not likely to be a panacea for climate change. Predicted irrigation requirements for soybean in the southeastern USA were increased 33 and 134% under the GISS and GFDL scenarios, respectively, based on simulations of Peart *et al.* and Curry *et al.* However, under the GFDL scenario, water resources would become scarce, and may not be readily available for crops. Some areas of the USA may have to adapt by irrigating less land area. Increasing temperatures and decreasing precipitation for the USA as predicted by the GFDL model would have a serious negative impact overall on agricultural productivity although producers in favourable regions may benefit from scarcity-mediated higher prices.

Assessing International Impacts of Rising CO_2 and Climate Change

Several assessments have been conducted on the impacts of rising CO_2 and global climatic changes on crop production patterns and economics responses within national or regional zones and in various countries around the world. Furthermore, progress has been made on predicting impacts of climate change using world trade models.

The studies conducted by Rosenzweig and colleagues used existing crop models and climate change scenarios from three GCMs (GISS, GFDL and United Kingdom Meteorological Office Model to predict changes in crop yields for a number of countries around the world. The doubled-CO_2 climate change scenarios (temperature, rainfall and evaporation changes) were based on the climate change potential expected from increases of all greenhouse-effect gases. These climate changes are expected to occur well before CO_2 concentration has actually doubled. Therefore, the CO_2 levels for fertilization effects were estimated to be 555 μ mol/mol rather than 660 μ mol/mol. Thus, the CO_2 fertilization effect photosynthetic ratios for four crops (soybean, wheat, rice and maize) were taken as 1.21, 1.17, 1.17 and 1.06, respectively.

The impact of climate change scenarios was more severe in the tropical latitudes than in the mid- or high-latitudes. For example, averaged over all three GCM scenarios, wheat production changes predicted in Brazil, India, China and Canada were -47, -43, -11 and -20%, respectively, without CO_2 effects, and -28, -13, +8 and +16% with CO_2 fertilization effects.

Simulations of soybean yields throughout the USA were a part of this international study. They found that aggregated yields were 2.42, 2.80, 2.37 and 1.31 Mg/ha for the BASE, GISS, GFDL and UKMO scenarios, respectively. Temperatures were about 4.0°C higher than BASE for the GISS and GFDL scenarios, but were about 5.2°C higher for the UKMO model.

Refinements and improvements in prediction methodology will continue, but these assessments provide the best currently available insights into the CO_2 fertilization and climate change effects on global crop productivity.

References

Vergara, B.S. 1980. Rice plant growth and development. In: *Rice Production and Utilization.* B.S. Luh (ed.). Avi Publishing, Westport, CT. pp. 75-86.

Wilson, C.A. and Mitchell, J.F.B. 1987. A doubled CO_2 climate sensitivity experiment with a global model including a simple ocean. *J. Geophys. Res.* 92: 13315-13343.

Wittwer, S.H. 1995. *Food, Climate, and Carbon Dioxide: The Global Environment and World Food Production.* CRC Press/Lewis Publishers, Boca Raton, Florida.

Woodwell, G.M. and Mackenzie, F.T. (eds.). 1995. *Biotic Feedbacks in the Global Climatic System: Will the Warming Increase the Warming?* Oxford University Press, New York.

Yoshida, S. 1977. Rice. In: *Ecophysiology of Tropical Crops.* P.T. Alvim and T.T. Kozlowski (eds.). Academic Press, London. pp. 57-87.

Yoshida, S. 1981. *Fundamentals of Rice Crop Science.* International Rice Research Institute, Los Baños, Philippines.

6

Climate Change and Crop Water Use

The projected climatic effects of the continuously increasing concentrations of CO_2 and other radiatively active trace gasses in the atmosphere have caused concern over the last decades and increasingly attracted scientists' and policy makers' attention. The expected changes at a global level will be reflected in changed weather conditions in the growing season at regional and local levels that directly affect agriculture and natural vegetation. The exchange of water and energy is determined both by the climate and by the gas exchange properties of the vegetation. Feedback on land cover, changed vegetation properties and local and regional climate is therefore of prime importance.

So far, the General Circulation Models (GCMs) are not able to give accurate predictions of climate change with great geographical detail. Also, feedback processes by the biosphere are still poorly represented in the models. These shortcomings limit predictions at the regional level that are needed in impact studies for agricultural production. Qualitative changes in potential production may well be estimated on the basis of the scenarios, but quantitative changes in the actual production, including the limitations caused by the changed variability of weather conditions (daily and seasonal amplitudes of temperatures, seasonal patterns in precipitation, cloudiness) remain difficult to estimate.

In addition to changes in precipitation and total water availability for irrigation, that directly affect agricultural production, changes in the pattern of water use by crop plants throughout the season may affect the outcome.

Of special concern is the change in the physiological functioning of the vegetation as a consequence of the changed atmospheric composition. Most plants react to the changed atmospheric CO_2 concentration with changed stomatal response, and not only is growth affected but also the transpiration. The complex nature of the physiological response in interaction with micrometeorological processes at the leaf and canopy level requires further attention.

Water Use and Water-use Efficiency

It should be realized that various expressions for water-use efficiency are used, and that the physiological and agronomic definitions may differ. Time-averaged water use and water-use efficiency (seasonal or periodic whole-plant transpiration rate (WPTR), and water-use efficiency (WUE)) are often confounded with instantaneous transpiration rate and transpiration efficiency (ITE). The latter expression is best described as the ratio between the instantaneous CO_2 fixation (actual net photosynthesis) and water loss by transpiration (both measured with a leaf or plant chamber). Water-use efficiency is the ratio of the net gain in dry matter over a given period, divided by the water loss (from the vegetation alone or from soil and vegetation together) over the same period. Mostly the ITE, normally only measured during the light period, is higher than WUE, both because of variations in ITE over day and season and because of carbon losses by respiration in the dark. WUE seems to be the more relevant parameter to the question of impact on agricultural production.

Interactions with Experimental Conditions

Most available data on water use and water-use efficiency have been collected from single plants grown in pots or hydroponically, and using porometer measurements supplemented with periodic weight determinations. Values for water-use efficiency and total water use in (semi-) field conditions over the full season, or an extended period, are scarce. Moreover, a complication for the correct estimation of the effect of elevated [CO_2] and temperature changes on transpiration is that almost all experiments in this field have been performed in environmentally controlled and generally well mixed and ventilated experimental set-ups (enclosures or Open-Top Chambers (OTCs)), where the indirect effects may not show up so prominently as in a real and outside future climate. Similar interpretation problems have been met in the evaluation of air pollution research.

Response at the level of stomata

Stomatal Density

The plant leaves lose water primarily by evaporation through the stomata. The stomatal density depends upon plant species, and can be related to the plant-ecotype (between 300 and 800 stomata/mm. Woodward correlated the decrease in the stomatal density over time, observed in herbarium leaves collected over the last centuries, with the rising CO_2 concentrations and concluded from the shift in d ^{13}C that the water-use efficiency concomitantly has improved. The nitrogen content in the leaves had dropped, in line with most data from elevated CO_2 experiments.

Experimentally, an increase in [CO_2] up to about 310 m l/l decreased the stomatal density, but sometimes no effect is found above this [CO_2]. This matter is still under dispute although such a correlation has also been confirmed for palaeo-records. Among species large differences in response of stomatal density to elevated [CO_2] seem to exist.

Experiments with a range of CO_2 concentrations (160-900 m l/l) showed an increase of stomatal density of rice and bean leaves, with a differential effect at abaxial (increasing) and adaxial sides. At subambient CO_2 concentrations stomatal density dropped. This is in contrast with the findings of Oberbauer *et al.* for tropical trees. The relative effect of changes in stomatal density and decreasing stomatal aperture at elevated [CO_2] for the water relations, however, have not been evaluated.

A gradual change in [CO_2] over the next century may lead to a natural selection that favours cultivars having a lower stomatal density, especially for water-limited growth conditions. It should, however, be realized that other environmental factors like salt stress can also modify stomatal density. Whatever the net effect, the resulting stomatal conductance is primarily determined by stomatal functioning, and much less by density.

Stomatal Activity

In the pathway from the stomatal cavity to the leaf surface, and from ambient air to the photosynthetic machinery in the mesophyll, the stomata are a major resistance for gas transport between the leaf and the surrounding air. A change in gas exchange resistance of the stomatal pores therefore affects the entrance of CO_2 and even more the exit of water vapour. The opening status of the stomata is a compromise between water loss and uptake of CO_2

from ambient air. In line with this proposition, stomatal response to elevated concentrations of CO_2 (C_a) in general is reflected in partial stomatal closure. The mechanism behind this stomatal closure is not yet clear.

The observations are in line with the notion that plants tend to regulate the internal CO_2 concentration (C_i) in the substomatal cavity such that for a given water vapour deficit there is a constant ratio (C_i/C_a) with the atmospheric concentration. Such a regulation would directly lead to the partial closure at elevated CO_2, as observed in many studies using porometers. Jackson *et al.* measured photosynthesis and water relations of native grassland species and calculated C_i/C_a for C_3 and C_4 plants. They confirmed the conservation of the value, with only a small (not significant) tendency to rise with increasing [CO_2].

The rate of photosynthesis and therefore the required supply of carbon dioxide is directly coupled to light intensity. In line with this, the conductance of stomata is also highly correlated to light. This relation can be modified by environmental conditions like drought or stress by air pollution. The ratio C_i/C_a in stationary conditions is about two-thirds for C_3 but about one-third for C_4 plants. The lower value for C_4 plants reflects the higher affinity for CO_2 of the C_4. photosynthetic pathway, and reflects the more efficient water use for these plants.

Acclimation of the Stomatal Movement

Little is known about the acclimation of the stomatal movement to long-term exposure to elevated [CO_2]. Some studies suggest that the lower stomatal conductance not only persists over extended exposure periods to elevated [CO_2], but is also conserved after subsequent lowering of the [CO_2] (Gorissen, pers. comm.). For glasshouse horticulture it was reported that the stomatal conductance of high-[CO_2] tomato plants was less sensitive to short-term [CO_2] fluctuations, and higher than that of plants grown at ambient concentrations and measured at double present [CO_2]. This may indicate that the sensitivity of stomatal movement of high [CO_2] plants to changes in [CO_2] is reduced, but still a lower stomatal conductivity exists when compared to leaves growing and measured in present [CO_2] conditions.

Effects at the Level of the Leaf

The partial closure of stomata is reflected in the reduced conductance at the leaf level. At a CO_2concentration double the present, conductance is reduced by 30-40%, although large differences among species exist.

Water loss by transpiration is not only affected by the conductivity of the stomata, but also by the driving forces for exchange of the water vapour from the leaf surface to the surrounding atmosphere. Therefore the gradient in partial pressure of the water vapour at the leaf surface is also of importance. All other factors being equal, the existing vapour pressure deficit (VPD) between stomatal cavity and surrounding air, the boundary layer, will increase at a reduced transpiration rate, and feed back to stimulate transpiration. The reduced transpiration will cool the leafless, and consequently a rise in temperature in the stomatal cavity and at the leaf surface may occur. Thus, in addition to the global greenhouse effect on air temperature, the temperature at the leaf surface may rise by 0.5 to 1.5 °C. In their FACE experiments (Free Air Carbon dioxide Enrichment) in Arizona, Kimball *et al.* measured an average rise in canopy temperature of 0.56°C over the growing season. Such a higher leaf temperature may also have important consequences for the longevity and photosynthetic capacity of the individual leaves and at the canopy level, as ageing may be accelerated. It is presently not known how different crops respond to the increase in leaf and canopy temperature.

The mostly negative feedback on water vapour exchange will at least partly counteract the effect of the reduced stomatal conductance even without rise in ambient temperature at the global scale. A rise of ambient growing season temperatures due to the greenhouse effect will also tend to stimulate transpiration. Depending on changes in air humidity it may affect VPD at the leaf surface determining the fluxes of sensible and latent heat and thereby the energy balance. Increased water availability may, however, offset the relative warming at the leaf surface, especially if, through preceding water saving and lower instantaneous requirements, partial stomatal closure alleviates water stress during periods of high water requirements in the diurnal sequence. It requires detailed microclimatological data and complex calculations to integrate these various mechanisms in operation at the leaf level.

Leaf Area and Weight

Changes in area of individual leaves also have a direct effect on water loss at a per leaf basis. In general a substantial increase in individual leaf weight is found. This is primarily due to an increased leaf thickness and additional accumulation of non-structural carbohydrates. Leith*et al.* found for soybeans that a small but significant increase in leaf area with CO_2 concentration may be expected (about 5% increase with a doubling [CO_2]). This is consistent with trends reported in other papers. In general, however, weight gain through increased leaf thickness is more important than that by the increase in leaf area. The relative gain in leaf area may also depend on growing conditions such as water shortage.

Effects at the Plant and Canopy Level

Aspects of the water economy at the leaf level also apply at the canopy level. Moreover, light distribution in the canopy, leaf age, humidity gradients and coupling with the atmosphere for different canopy layers and canopy structures all have their effect. The direct effects on vegetation properties discussed here are linked to the change in CO_2 concentration and temperature, but indirect effects of a climate change related with changes in ambient temperature and air humidity, coupling to the lower atmosphere and albedo changes, may even dominate. Mismanagement or over-exploitation of natural resources may lead to changes in the vegetation and even to desertification, and via a positive feedback aggravate changes in local and regional climate as studied in the EU-EFEDA-programme (European Field Experiment in Desertification-threatened Areas). This field of research is still under development and the subject of international programmes like the IGBP-BAHC (International Geosphere-Biosphere Programme - Biospheric Aspects of the Hydrological Cycle).

The leaf area index (leaf surface area per unit soil surface area) of a crop with adequate water supply at elevated [CO_2] increases, especially early in the season, as a result of earlier and more rapid leaf production in the vegetative growth phase. This applies especially for indeterminate growing species and under non-limiting supply of nutrients. Early development of the canopy will lead to an earlier full ground cover, and may thus limit water loss from direct soil evaporation. Depending on the local precipitation and available soil water reserves, such an early enhanced canopy development may also be favourable for the full utilization of water resources. The higher transpiration early in the season may also lead to an earlier depletion of water reserves in the soil.

In their FACE experiments Hendrey *et al.* found that for cotton leaf area increased by 25% in dry plots, but by only 11% in irrigated plots as compared to the respective controls. Such enhancements in leaf area and biomass formation in dry conditions could improve the duration of the vegetative soil cover, and help counteract erosion and other forms of land degradation.

Palnt Water Status

It has been shown that plant water status is generally improved at elevated CO_2. Part of the effect can be ascribed to a reduced transpiration demand (per unit leaf area), and therefore a partial alleviation of the water stress. Also for plants with a normal water supply, pre-dawn and midday water potentials have been found to be less negative in high [CO_2] Even in saline conditions the effect of partial stomatal closure is reflected in a higher salt tolerance. As a consequence, the restriction of photosynthesis under low water supply may be less severe, or be delayed.

Changes in Rooting Pattern

Several studies have reported an increased input of carbon in below-ground processes. An important aspect to be considered in reviewing the literature at this point is that most experiments concern plant growth over a short period and with a soil volume limited by pot or container size. In a study with container-grown grasses and clover Nijs *et al.* showed that in a rapidly developing terminal drought period water-use efficiency is about doubled, and stress is developing later under elevated [CO_2]. These conditions are thus improving the possibility to escape drought stress. Even though the trend is clear, elevated [CO_2] plants grown with restricted soil volume may not take full advantage of a larger root system.

In (semi-) field experiments the larger root system indeed leads to a better exploration of the soil volume, and an earlier, or higher, rooting density at the larger depths. Experiments with potted plants can therefore not be expected to show the full extent to which elevated [CO_2] plants may profit in a (transient) drought period. The observation of Morison and Gifford that the pattern of depletion of soil water reserves is rather similar may partly be an underestimation of the real moisture availability in field conditions because of a better access to deeper soil layers. It has also been observed that a faster recovery of the crop is possible when grown at elevated [CO_2]. The explanation of this observation has not yet been given, but could be related to deeper and more extensive rooting patterns.

Temperature Effects

The importance of temperature at the canopy level is two-fold. Firstly higher temperatures increase transpiration by changing the VPD at the leaf surface, and secondly the higher canopy temperature may lead to an accelerated ageing of the foliage, and a shortening of the growing season, or, for example, grain-filling period. The latter type of effect is discussed extensively in crop growth modelling studies that describe changes in crop productivity due to climate change.

On theoretical grounds it can be expected that transpiration losses will increase with higher air temperatures. It is likely that the evaporative demand as determined by the vapour pressure deficit would increase by about 5 to 6% per degree warming. The overall effect of a higher temperature alone is an intensification of the hydrological cycle. It should be realized that the combined result of a higher ambient temperature, leading to a higher evaporative demand, and partial stomatal closure, counteracting this, could be overridden by changes in other environmental and atmospheric conditions like soil water availability, precipitation patterns, cloudiness and air humidity.

Effects over the Growing Season

The effects of elevated CO_2 alone on transpiration integrated over the season show that any of the cases (higher water use; no change; lower water use) can be found. Especially in the case of non-limiting water supply a higher water use can be expected even with a lower stomatal conductance as leaf area will increase with about the same factor as total biomass. This will compensate for the reduction in water exchange per leaf area. This applies especially in early season. In conditions with transient water shortage, profit from partial stomatal closure will be highest, as both the depletion of soil water reserves is delayed and plant sensitivity to midday water shortage (suppressed photosynthesis) is lessened.

In conditions with adequate water supply, water use over the whole season, and especially in periods in the season with a closed canopy (e.g., with leaf area index over 3), is little different if at all. ITE and WUE (24h basis) are both higher (about 40-50%) at elevated [CO_2], but transpiration per unit soil surface is not changed significantly. Here treatment temperatures have been kept unchanged, tracking outside conditions. The improvement in seasonal water-use efficiency (SWUE) ranges from 23% (wheat) to 54% (faba bean), similar to the results reported by Dijkstra *et al.*

Similarly, Baker and Allen presented results of canopy gas exchange for soybean and citrus at a range of temperatures. Although day time water-use efficiency (DWUE) fell with temperature, they measured at 800 m l/l [CO_2] a doubling (soybean) or more of the DWUE, independent of temperature (28-35°C). Seasonal water-use efficiency varies with crop and growth conditions

Water-use efficiency rises especially in conditions with water limitations. In their study of the potential changes of productivity of cool-season legumes, Grashoff *et al.* concluded that for rainfed faba bean crops in the Netherlands, Syria and Israel productivity would increase, especially under water limitation. Faba bean is a special case, because crop development rate is accelerated by drought. A small increase in temperature (1.7°C) would decrease yield, but the lower water requirement at higher [CO_2] (1.7°C and 460 ml/l for the climate in 2030) more than compensates for this, raising yields from 2.8 to 4.7 t/ha (Syria), 3.9 to 5.0 t/ha (Israel) and 5.1 to 5.7 t/ha (Netherlands). In fully irrigated conditions yields would be higher, but the relative CO_2 effect would be lower. Interestingly the standard deviation of the predicted yield (>10 year simulated average) shows a tendency to decrease in the changed climate scenario.

Estimation of Changes in the Water Use of Arable Crops

One of the more complex aspects of the estimation of changes in the water use of arable crops is the extrapolation from leaf and canopy to the field and region. This has long been recognized by the international scientific community, and has given rise to the establishment of several large-scale measuring campaigns (HAPEX: Hydrologic Atmospheric Pilot Experiment (Sahel and Spain), GEWEX: Global Energy and Water Cycle Experiment (World Climate Research Programme)) and to the start of the IGBP-BAHC programme. A major part of the efforts will go into the development of the SVAT-models (Soil-Vegetation-Atmosphere-Transfer models). The extrapolation from the stomatal and leaf level to the region have been the subject of experimental and theoretical studies. In general the structure of the canopy determines to a large extent the transfer of heat and water vapour to the lower atmosphere. Most models to date neglect or oversimplify the feedback between the vegetated surface and the lower atmosphere, the planetary boundary layer (PBL). Jacobs showed in a study using coupled models (one-dimensional model for the PBL coupled to a modified Penman-Monteith big-leaf model) that in such a model the sensitivity of regional transpiration to changes in surface resistance is reduced (about halved) but that the sensitivity to changes in albedo is increased by between 25 and 250% relative to the values obtained without PBL feedback.

In addition to the sensitivity of stomata to photosynthesis and [CO_2], Jacobs introduced an additional correlative relationship between the humidity deficit at the leaf surface and the C_i/C_a ratio. The changes in transpiration caused by changes in surface resistance are damped. However, within the canopy the changes in the specific humidity lead to a positive feedback, e.g., the CO_2 effects on surface resistance are enhanced. The inclusion of PBL feedback and stomatal response to humidity deficit leads in his study to an estimated overall decrease of the regional transpiration by 10 to 30%. Modifying factors are surface roughness determined by the vegetation type, temperature and air humidity. In particular, information about the last factor, extrapolated from regionalized GCM output, is virtually absent or highly speculative.

Arable crops, especially in the temperate regions in general, have a low surface roughness, and in that case the transpiration is primarily determined by the radiation energy. This is also true for pastures. In rangelands with sparse higher vegetation the higher roughness leads to a stronger coupling, a situation comparable to forests and mixed vegetation.

Present knowledge does not allow a firm statement to be made concerning a future situation with respect to water-limited agricultural production. At a regional production level the available quantitative data from regionalized GCM runs are insufficient to be used as direct inputs for crop growth models. At a global level the simultaneous changes in rainfall pattern, air humidity and possible shifts in vegetation zones add to uncertainty, as many and often non-linear feedbacks are expected to operate.

The limited amount of available experimental data has as a trend that water use per unit soil surface area will change little (-10 to +10%). As, however, the general trend towards an improved water-use efficiency is clear, the productivity per unit of available water is expected to rise by 20-40%, probably much less than the value (100%) calculated from a reduced stomatal conductivity and an increased photosynthesis. Some studies show that in situations with marginal water availability the threshold for a successful crop may shift to lower values. Whether at otherwise unchanged water availability this would open up possibilities to reverse existing trends towards desertification in certain areas is very questionable, as apart from precipitation falling short to maintain existing vegetation, other aspects like over-exploitation might dominate. It should, however, be emphasized that present knowledge on feedback among vegetation characteristics, gas exchange and albedo and the regional climate is insufficient to draw firm conclusions.

Some related problems have not been touched upon. For example, changes in precipitation distribution throughout the season may lead to shifts in the accessibility of fields for farm operations in early spring or late autumn. In addition they may cause changes in soil moisture and thereby modify the rate of mineralization of nutrients (nitrogen, phosphorus). Such interactions, combined with changes in (air and soil) temperatures, may markedly change soil fertility and thereby local farming systems and crop productivity. Depending on the aim of the study, such impacts have to be considered more or less relevant to the present subject.

In many studies the impact of climate change on crop growth and yield is analysed using crop simulation models. A proper analysis of the performance of such models should be made to verify their reliability in the projected conditions of changed atmospheric composition and changed climate. The absolute values of the regionally and globally aggregated predicted crop yields, often calculated using a proportionality factor for the CO_2response, are at present probably less reliable than the predicted sensitivity of the yield to various climate and management factors.

The above considerations relate to a projected world, changed primarily in terms of climate and atmospheric composition. It should be emphasized that at many points agricultural practice is very dynamic, and will respond to changed conditions by adaptation. Crop and cultivar choice will, in most instances and in the most productive areas, change over time and gradually incorporate the traits necessary for adapted performance, or change to better adapted species. A rising temperature will for most current cultivars, for instance, accelerate crop development. This would in itself lead to a reduced water use over the shortened growth period, but also to a loss of potential yield. One may argue that farmers will repair such a loss of production potential by a proper choice of adapted cultivars or crop species, unless temperatures exceed the appropriate temperature window.

These developments require, however, that new technologies and genetic resources be practically and economically accessible for all farmers, a situation that is not reached at present for farmers in arid and semi-arid regions in developing countries where the risks but probably also opportunities for agricultural production may be greatest.

References

Jacobs, C.M.J. 1994. *Direct Impact of Atmospheric CO_2 Enrichment on Regional Transpiration.* Thesis. Wageningen Agricultural University. 179 p.

Kimball, B.A., Mauney, J.R., Nakayama, F.S. and Idso, S.B. 1993. Effects of increasing atmospheric CO_2 on vegetation. *Vegetatio*104/105: 65-75.

McNaughton, K.G. and Jarvis, P.G. 1991. Effects of spatial scale on stomatal control of transpiration. *Agric. For. Meteorol.* 54: 279-301.

Morison, J.I.L. 1987. Intercellular CO_2 concentration and stomatal response to CO_2. In: *Stomatal Function.* E. Zeiger, G.D. Farquhar and I.R. Cowan (eds.). Stanford Univ. Press, California. pp. 229-251.

7

Effects of Global Warming on Plan Growth and Yields

Gaseous emissions from human activities are substantially increasing the concentrations of atmospheric greenhouse gases, particularly carbon dioxide, methane, chlorofluorocarbons and nitrous oxides. Global circulation models predict that these increased concentrations of greenhouse gases will increase average world temperature. Under the business-as-usual scenario of the Intergovernmental Panel on Climate Change (IPCC), global mean temperatures will rise 0.3°C per decade during the next century with an uncertainty of 0.2 to 0.5%.

Thus global mean temperatures should be 1°C above the present values by 2025 and 3°C above the present value by 2100. Although global circulation models do not all agree as to the magnitude, most predict greenhouse warming. There is also general agreement that global warming will be greater at higher latitudes than in the tropics. Different global circulation models have predicted that global warming effects will vary diurnally, seasonally and with altitude.

It is also possible that there will be an autocatalytic component to global warming. Photosynthesis and respiration of plants and microbes increase with temperature, especially in temperate latitudes. As respiration increases more with increased temperature than does photosynthesis, global warming is likely to increase the flux of carbon dioxide to the atmosphere which would constitute a positive feedback to global warming.

Mechanisms for Heat Tolerance

Crop plants are immobile. They must adapt to prevalent soil and weather conditions. Except for transpirational cooling, plants are unable to adjust their tissue temperatures to any significant extent. On the other hand, plants have evolved several mechanisms that enable them to tolerate higher temperatures. These adaptive thermotolerant mechanisms reflect the environment in which a species has evolved and they largely dictate the environment where a crop may be grown.

Four major aspects of thermotolerance have been studied: (1) thermal dependence at the biochemical and metabolic levels; (2) thermal tolerance in relation to membrane stability; (3) induced thermotolerance through gradual temperature increase vis-a-vis production of heat shock proteins; and (4) photosynthesis and productivity during high temperature stress.

Biochemical Processes

Temperature effects on the rates of biochemical reactions may be modelled as the product of two functions, an exponentially increasing rate of the forward reaction and an exponential decay resulting from enzyme denaturation as temperatures increase. The greatest concern is whether it is possible to increase the upper limit of enzyme stability to prevent denaturation.

Failure of only one critical enzyme system can cause death of an organism. This fact may explain why most crop species survive sustained high temperatures up to a relatively narrow range, 40 to 45°C. The relationship between the thermal environment for an organism and the thermal dependence of enzymes has been well established.

The shape of this function also describes temperature effects on most biological functions, including plant growth and development. The function can be categorized by the three cardinal temperatures - minimum, optimum and maximum. Modellers frequently simplify the relationship into a stepwise linear function. The stepwise linear function has a plateau rather than an optimum temperature.

The thermal dependence of the apparent reaction rate for selected enzymes may indicate the optimal thermal range for a plant. The range over which the apparent Michaelis-Menten constant for CO_2 (K_m) is minimal and stable is termed the thermal kinetic window. For crop plants, the thermal kinetic window (TKW) is generally established as a result of thermally induced lipid phase changes, rubisco activity and the starch synthesis pathway in leaves and reproductive organs.

In cotton and wheat, the time during which foliage temperature remained within the TKW was related to dry matter accumulation. The cumulative time that rainfed crop foliage is outside the TKW provides an index of the degree of extreme temperature stress of the environment. Irrigation is one management option to reduce crop exposure to heat stress.

Temperatures that inhibit cellular metabolism and growth for a cool season C_3 species such as wheat may not inhibit warm-season C_3species such as rice (*Oryza sativa* L.) and C_4 species such as sorghum, maize (*Zea mays* L.) and sugar cane (*Saccharum spontaneum*spp.). The identification of TKWs for different species can aid in the interpretation of the differential temperature stress responses for crop growth and development among species.

Thermal Stability of Cell Membranes

The plasmalemma and membranes of cell organelles play a vital role in the functioning of cells. Any adverse effect of temperature stress on the membranes leads to disruption of cellular activity or death. Heat injury to the plasmalemma may be measured by ion leakage. Injury to membranes from a sudden heat stress event may result from either denaturation of the membrane proteins or from melting of membrane lipids which leads to membrane rupture and loss of cellular contents.

Heat stress may be an oxidative stress. Peroxidation of membrane lipids has been observed at high temperatures, which is a symptom of cellular injury. Enhanced synthesis of an anti-oxidant by plant tissues may increase cell tolerance to heat but no such anti-oxidant has been positively identified.

A relationship between lipid composition and incubation temperature has been shown for algae, fungi and higher plants. In *Arabiodopsis,*exposed to high temperatures, total lipid content decreases to about one-half and the ratio of unsaturated to saturated fatty acids decreases to one-third of the levels at temperatures within the TKW. Increase in saturated fatty acids of membranes increases their melting temperature and thus confers heat tolerance. An *Arabiodopsis* mutant, deficient in activity of chloroplast fatty acid W-9 desaturase, accumulates large amounts of 16:0 fatty acids, resulting in greater saturation of chloroplast lipids. This increases the optimum growth temperature.In cotton, however, heat tolerance does not correlate with degree of lipid saturation and similar differences in genotypic differences in heat tolerance have been unrelated to membrane lipid saturation in other species. In such species, a factor other than membrane stability may be limiting growth at high temperature.

Heat Shock Proteins

Synthesis and accumulation of proteins were ascertained during a rapid heat stress. These were designated as 'Heat Shock Proteins' (HSPs). Subsequently it was shown that increased production of these proteins also occurs when plants experience a gradual increase in temperature more typical of that experienced in a natural environment.

Three classes of proteins as distinguished by molecular weight account for most HSPs, namely HSP90, HSP70, and low molecular weight proteins of 15 to 30 kDa (LMW HSP). The proportions of the three classes differ among species. In general, heat shock proteins are induced by heat stress at any stage of development. Under maximum heat stress conditions, HSP70 and HSP90 mRNAs can increase ten-fold and LMW HSP increase as much as 200-fold. Three other proteins, though less important, are also considered to be heat shock proteins viz. 110 kDa polypeptides, ubiquitin, and GroEL proteins.

In arid and semi-arid regions, dryland crops may synthesize and accumulate substantial levels of heat shock proteins in response to elevated leaf temperatures. The induction temperature for synthesis and accumulation of heat shock proteins in laboratory-grown cotton ranged from 38 to 41°C. Soil water deficits resulting in midday canopy temperature of 40°C or greater for two to three weeks were used to study heat shock proteins in field-grown cotton. A comparison of polypeptide patterns of dryland and irrigated cotton leaves showed that at least eight new polypeptides accumulated in about half of the dryland leaves analysed. The polypeptides that accumulated in the dryland leaves but not irrigated cotton leaves had molecular weights of 100, 94, 89, 75, 60, 58 and 21 kDa. In a similar experiment with field-grown soybean, several heat shock proteins were observed in both irrigated and dryland treatments, although levels were greater in the non-irrigated treatments.

Correlation between synthesis and accumulation of heat shock proteins and heat tolerance suggests, but does not prove, that the two are causally related. Further evidence for a causal relationship is that some cultivar differences in heat shock protein expression correlate with differences in thermotolerance. In genetic experiments, heat shock protein expression co-segregates with heat tolerance. Another evidence for the protective role of heat shock protein is that mutants unable to synthesize heat shock proteins, and cells in which HSP70 synthesis is blocked or inactivated, are more susceptible to heat injury.

The mechanism by which heat shock proteins contribute to heat tolerance is still not certain. One hypothesis is that HSP70 participates in ATP-dependent protein unfolding or assembly/disassembly reactions and that they prevent protein denaturation during stress. If this mechanism is true, then heat shock proteins may provide a significant basis for increasing heat tolerance of crop plants in a global warming situation. The LMW HSPs may play a structural role in maintaining cell membrane integrity during stress. Other heat shock proteins have been associated with particular organelles such as chloroplasts, ribosomes and mitochondria. In tomato (*Lycopersicon esculentum* L.), heat shock proteins aggregate into a granular structure in the cytoplasm, possibly protecting the machinery of protein synthesis.

HSPs provide a significant opportunity to increase heat tolerance of crops. To elucidate their mechanisms of action and to exploit their potential contribution to increasing heat tolerance, four lines of investigations are suggested:

1. Establish the biochemical activities of individual HSPs as a preliminary step.
2. Characterize the genetic variability of specific heat shock proteins across a wide range of germplasm. Develop iso-population and near isogenic lines selected for production of low and high levels of HSP synthesis.
3. As HSPs appear to participate in maintaining the conformation or assembly of other protein structures, analyse the molecular details of these processes and establish all participating protein substrates. Such biochemical studies are needed to understand how these processes protect or allow recovery from heat stress.
4. Identify specific HSP mutants or create transgenic mutant plants to complement molecular and biochemical understanding with genetic approaches.

Photosynthesis and High Temperature Stress

Variability in leaf photosynthetic rates within or between species is often unrelated to differences in productivity. Similarly, high photosynthetic rates at high temperatures do not necessarily support high rates of crop dry matter accumulation. The temperature optimum for photosynthesis is broad, presumably because crop plants have adapted to a relatively wide range of thermal environments. A 1 to 2°C increase in average temperature is not likely to have a substantial impact on leaf photosynthetic rates. Further, there is a possibility that photosynthesis of crop plants can adapt to a slow increase in global average temperatures. Thus, global warming is not likely to affect photosynthetic rates per unit leaf area gradually or on a closed canopy basis over the next century.

While photosynthetic rates were found to be temperature-sensitive in other crops, wheat and rice appear to be different. In wheat, no measurable differences were found in photosynthetic rates per unit flag leaf area or on a whole-plant basis in the temperature range from 15 to 35°C. In rice, there is little temperature effect on leaf carbon dioxide assimilation from 20 to 40°C.

Recent research has shown significant variation among wheat cultivars with respect to reduction in photosynthesis at very high temperature. Photosynthesis of germplasm adapted to higher temperature environments was less sensitive to high temperature than was germplasm from cooler environments. When this germplasm was grown under moderate (22/17°C) and high (32/27°C) temperatures in the seedling stage or from anthesis to maturity, there was a highly significant correlation between photosynthesis rate and either seedling biomass (r=0.943***) or grain yield of mature plants (r=0.807**). Genotypes most tolerant to high temperatures had the most stable leaf photosynthetic rates across temperature regimes or they had the longest duration of leaf photosynthetic activity after anthesis and high grain weights. The above relationship was exemplified by 'Ventnor' from the high temperature area of Australia and 'Lancero' from the high altitude area of Chile.

Despite observed negative effects of high temperature on leaf photosynthesis, the temperature optimum for net photosynthesis is likely to increase with elevated levels of atmospheric carbon dioxide. Several studies have concluded that CO_2-induced increases in crop yields are much more probable in warm than in cool environments. Thus, global warming may not greatly affect overall net photosynthesis.

Night Respiration Response to Co_2 and Temperature

In tomato (*Lycopersicon esculentum*) and cotton (*Gossypium hirsutum*), elevated CO_2increased dark respiration possibly because of increased carbohydrate accumulation in tissues. The latter has been shown to increase alternative pathway respiration as well.

Apparent dark respiration may decline under elevated CO_2 if there is dark CO_2 fixation or if elevated CO_2 directly inhibits or inactivates respiratory enzymes as may occur through increased formation or carbamate

Crop Growth and Development

Wheat development is customarily divided into vegetative and reproductive phases, with either ear emergence or anthesis as the event that separates the two phases. In the past 30 to 40 years, the sequence of pre-anthesis phenological events has been critically assessed with respect to grain yield potential and sensitivity to weather variables, particularly prevailing temperature and day length. Several systems are used to classify the sequence of phenological events. We identify five such developmental stages:

(i) germination - seeding to seedling emergence;

(ii) canopy development - emergence to first spikelet initiation, the double ridge stage;

(iii) spikelet production - first spikelet initiation to terminal spikelet formation;

(iv) spikelet development - terminal spikelet formation to anthesis;

(v) grain development - anthesis to maturity.

These stages are generally based on early recognized features of the apical meristem. They mark significant changes in morphology or physiology of different crop organs. Numbers of leaf and tiller primordia are determined before spikelet initiation but their subsequent growth and development are controlled by temperature and day length during the differentiation of spikes into spikelets. Similarly, floret number within each spikelet is established by anthesis, at which time the potential grain number per spike is established.

Productivity of wheat and other crop species falls markedly at high temperatures. Wheat in India is invariably exposed to extreme temperatures during some stages of development. In Australia, wheat is usually exposed to brief periods of heat stress during grain development.

All stages of development are sensitive to temperature. It is the main factor controlling the rate of crop development. Development generally accelerates as temperature increases, a phenomenon that is often described as a linear function of daily average temperature. The growing degree day concept is a common example of a linear model of developmental response to temperature

Vegetative Phase

Several experiments have observed the effects of temperature on the duration from sowing or emergence to heading under controlled environment and field conditions. Unfortunately, however, few experiments have been conducted with enough cultivars to assess the genetic variability in this trait. The major conclusions from these studies are:

1. All genotypes are sensitive to temperature at one stage or another. Temperature sensitivity, however, varies greatly with genotype.
2. Phenological stages differ in sensitivity to temperature.
3. The duration of phase from sowing to first spikelet initiation is less sensitive to change in temperature than are other phases, although genotypes do differ in thermotolerance during this phase.
4. The stages during which environment has the greatest impact on yield are from first spikelet initiation or terminal spikelet formation until anthesis. Spikelet number and floral number (potential grain number), both dominant yield contributing attributes, are established during these phases. Grain weight, on the other hand, appears to be much less sensitive to heat stress than is grain number.

Grain Development Phase

In experiments under controlled conditions from 25 to 35°C, mean grain weight declined 16% for each 5°C increase in temperature. In pot experiments, grain yield decreased by 17% for each 5°C rise. For every 1°C rise in temperature, there is a depression in grain yield by 8 to 10%, mediated through 5 to 6% fewer grains and 3 to 4% smaller grain weight.

To elucidate the causal factor for reduced grain filling in wheat because of higher temperatures, Wardlaw, studied the three main components of the plant system. The three components are: (a) source - flag leaf blade; (b) sink - ear; and (c) transport pathway - peduncle. He observed that photosynthesis had a broad temperature optimum from 20 to 30°C with photosynthesis declining rapidly at temperatures above 30°C. The rate of 14C assimilate movement out of the flag leaf, phloem loading, was optimum around 30°C; the rate of 14C assimilate movement through the stem was independent of temperature from 1 to 50°C. Thus, in wheat, temperature effects on translocation result indirectly from direct temperature effects on source and sink activities.

In a subsequent experiment with source-sink relationships altered through grain excision, defoliation and shading treatments, heat stress still reduced grain weight. This result supports the earlier findings that temperature effects on grain weight are direct effects rather than assimilate availability. Furthermore, respiration effects do not appear to be the direct cause of decreased grain size in heat-stressed wheat.

Reduction of grain weight by heat stress may be explained mostly by effects of temperature on rate and duration of grain growth. As temperature increased from 15/10°C to 21/16°C, duration of grain filling was reduced from 60 to 36 days and grain growth rate increased from 0.73 to 1.49 mg/grain/day with a result of minimal influence on grain weight at maturity. Further increase in temperature from 21/16°C to 30/25°C resulted in decline in grain filling during 36 to 22 days with a minimal increase in grain growth rate from 1.49 to 1.51 mg/grain/day. Thus, mature grain weight was significantly reduced at the highest temperature.

Research on the effects of brief periods of ear warming after anthesis on ear metabolism have identified differential responses of starch and nitrogen accumulation in grain of four wheat cultivars. Warming increased the rate of dry matter accumulation in all the cultivars but the increase was less in cv. Aus 22645 than in the other cultivars studied. Rate of increase in nitrogen accumulation was, however, higher than the increase in total dry matter accumulation. Under long-term exposure to heat stress, increased grain nitrogen concentration is almost entirely as a result of decreased starch content rather than a change in total grain quality. The conversion of sucrose to starch within the endosperm is decreased by elevated temperatures. Furthermore, heat stress effects on final grain weight were associated with reduced levels of soluble starch synthetase activity.

In summary, high temperature reduction of grain yield results from: (a) reduced numbers of grains formed; (b) shorter grain growth duration; and (c) inhibition of sucrose assimilation in grains.

Extreme Temperature Effects on Crops

There are two major forms of extreme temperature stress on crops - heat and cold. An increase in global temperatures may have either or both of these two acute effects: more frequent high temperature stress and less frequent cold temperature stress.

Increase in temperature will lengthen the effective growing season in areas where agricultural potential is currently limited by cold temperature stress. Thus, increased temperature will cause a poleward shift of the thermal limits to agriculture. This poleward shift will be especially important for crops such as rice that have tropical centres of origin and adaptation but are also grown in temperate latitudes during warm seasons. Global warming impact will be greater in the northern than southern hemisphere because there is more high-latitude area cultivated in the northern hemisphere.

Plants were grown at 21/16°C and some (W) were warmed, between 10 and 20 days after anthesis, to 33/25°C and then returned to 21/16°C where they stayed until maturity. Control plants (C) were grown at 21/16°C throughout. Values given are the means ±.e; values in parentheses are percentage of cotton values. (From Bhullar and Jennar,; reproduced with permission.)

Increased temperature would also affect the crop calendar in tropical regions. In the tropics, however, global warming, though predicted to be of only small magnitude, is likely to reduce the length of the effective growing season, particularly where more than one crop per year is grown. In semi-arid regions and other agro-ecological zones where there is wide diurnal temperature variation, relatively small changes in mean annual temperatures could markedly increase the frequency of highest temperature injury. For example, canopy temperature is 10 to 15°C higher in dryland cotton (*Gossypium hirsutum* L.) than in irrigated cotton. Thus, global warming would reduce dry matter accumulation in dryland cotton because of increased respiration, and reduced photosynthesis and cellular energy.

In India, the growing season for wheat is limited by high temperatures at sowing and during maturation. As wheat is grown over a wide range of latitudes, it is frequently exposed to temperatures above the threshold for heat stress. For example, rainfed wheat depends on soil moisture remaining after the monsoon rains recede in September. High maximum and minimum temperatures in September (about 34/20°C), which adversely affect seedling establishment, accelerate early vegetative development, reduce canopy cover, tillering, spike size and yield. Hence, sowing is typically delayed until after mid-October when seedbeds have cooled, though much of the residual soil moisture may be lost. High temperatures in the second half of February (25/10°C), March (30/13°C) and April (30/20°C) reduce the numbers of viable florets and the grain-filling duration. High temperature stress particularly reduces yield of wheat sown in December/January which is necessitated in some regions because of the multiple cropping system.

The situation is similar for sorghum (*Sorghum bicolor* (L.) Moench) and pearl millet (*Pennisetum glaucum* (L.) R.Br.) which are exposed to extreme high temperatures in Rajasthan, India. After sowing, air and soil temperatures often exceed 40°C and midday soil surface temperatures above 50°C are common.

Acute effects of high temperature are most striking when heat stress occurs during anthesis. In rice, heat stress at anthesis prevents anther dehiscence and pollen shed, to reduce pollination and grain numbers.

Clearly, many crops in tropical areas are already subjected to heat stress. If temperatures increase further, crop failure in some traditional areas would become more commonplace.

Long-term Effects of High Temperatures on Crops

More important than acute effects of extreme temperature stress are the chronic effects of continuously warmer temperatures on crop growth and development. Chronic effects of high temperature include effects on grain growth discussed above. Record crop yields clearly reflect the importance of season-long effects on crop yields: crops generally yield the most where temperatures are cool during growth of the harvested component.

Crop growth simulations show that rice yields decrease 9% for each 1°C increase in seasonal average temperature. This chronic effect of high temperature differs significantly from the acute effect of short-term temperature events, because seasonal temperature effects are mostly a result of effects on crop development. For most grain crops, there is much greater genotypic variation in thermal requirements for vegetative than for reproductive development. As long-term temperatures increase, grain-filling periods decrease, and there appears to be little scope to manipulate this effect through existing genetic variation within species.

Effects of Elevated Surface-level UV-B Radiation or O_3 on Crops

Surface-level ultraviolet (UV)-B radiation (280-320 nm) and ozone (O_3) are components of the global climate and any increases in their levels can lead to adverse effects on crop growth and productivity on a broad geographic scale. Possible increases in surface UV-B radiation are attributed to the depletion of the beneficial stratospheric O_3 layer. On the other hand, increases in surface-level O_3 that in many regions are largely the result of photochemical oxidant pollution, are also part of the general increase in the concentrations of the so-called 'greenhouse' gases (e.g., carbon dioxide, CO_2; methane, CH_4; nitrous oxide, N_2O; chlorofluorocarbons, CFCs) that may lead to global warming. In the context of climate change, it is therefore important to maintain a holistic view and recognize that UV-B and O_3 levels at the surface are only parts of the overall system of atmospheric processes and their products.

In recent years there have been a number of technical reviews or assessments of the direct effects of elevated surface-level UV-B or O_3 on crops and other terrestrial.

In their analysis of the UV-B exposure studies, Krupa and Kickert, noted that very different crop responses had been observed for the same crop species by different investigators at different times and locations. In many cases, these differences reflect cultivar and varietal differences in sensitivity within a given species. Another reason for the differences is probably from the use of different UV-B lamps, exposure systems and action spectra for computing biologically effective UV-B flux densities. Action spectra play a key role in the understanding of biological impacts of UV-B because of: (1) the differential sensitivities of various responses across the range of wavelengths in the UV-B region, and (2) the magnitudes of the differences in surface flux densities at these wavelengths resulting from the shape of the O_3 absorption spectrum.

The results obtained with a given species are frequently contradictory when comparing the effects of UV-B exposures in growth chambers or greenhouses to those under field conditions. Such differences may well arise not only from differences in the microclimatic radiant and heat energy budgets extant in the different exposure systems at the times of exposure, but also because of differences induced by the environmental conditions under which the plants were grown beforehand. For example, although the cuticle may act as a barrier to UV-B, greenhouse-grown plants are known to have a much thinner and less well-developed cuticle than field-grown plants and thus might exhibit greater sensitivity.

However, probably the factor of greatest importance in determining the relevance of many growth chamber and greenhouse studies is the intensity of Photosynthetically Active Radiation (PAR) to which the plants were exposed. For example, Biggs*et al.*, acknowledged that the sensitivities to UV-B of the soybean cultivars they studied were enhanced by the low (one-eighth of full sunlight) PAR levels used, because of the minimal activity of photorepair processes.

The results of field studies using selective filters to remove UV-B or filter/UV-lamp combinations are themselves far from conclusive. For example, Becwar *et al.*, found no significant effects on dry matter accumulation of pea, potato, radish and wheat plants grown for 50 days at a high elevation site (3000 m) in Colorado with high PAR and UV-B fluxes, even when filtered lamps increased the effective UV-B radiation by 52%. The only significant effect observed was an early slight decrease in wheat plant height growth that had disappeared by the time of final harvest. In contrast, Teramura *et al.*, reported net adverse effects of enhanced UV-B on the yield of the sensitive soybean cultivar, Essex, based on a six-year study, both for individual years and when averaged over the years. However, this effect was only observed at the higher UV-B enhancement used (computed to be equivalent to a 25% stratospheric O_3depletion). There was no adverse effect on the cultivar, Williams, when averaged over the six years; indeed a lower UV-B enhancement (equivalent to 16% O_3 depletion) resulted in a significant average increase in yield. The authors attributed the wide range of responses observed for either cultivar over the years to a strong influence of seasonal microclimate.

Recent studies at the University of Lancaster have used growth chambers in which UV-B enhancement was provided at high overall light intensities approximating two-thirds full sunlight (N. Paul and A.R. Wellburn, pers. comm.). Under these artificial conditions that approach those typical of the field with respect to light intensities, the vegetative growth of pea plants and their rates of photosynthesis were found to be unaffected by increased UV-B levels. However, yields were found to be somewhat depressed presumably because of adverse effects of UV-B at various stages in the process of sexual reproduction. The vegetative growth of a range of barley cultivars was unaffected by increased UV-B flux, although the cultivar, Scout, that contains little if any flavonoids, suffered visible injury. Such observations tend to confirm the conclusions of others. For example, Beyschlag *et al.*, found no adverse effect of UV-B on the photosynthetic rates of competing wheat and wild oat plants, although the competitiveness of wheat was increased because of UV-B-induced inhibition of the height growth of the wild oat plants. Increased flavonoid production has long been hypothesized as a protective feature of many species and cultivars.

As noted in the case of UV-B effects, there are also appreciable differences among species and cultivars with regard to their response to O_3. Such differences in sensitivity exist with respect to both growth responses and the induction of the visible symptoms of acute injury. Seasonal and locational differences also contribute to the variability of response, as illustrated in many of the extensive field studies undertaken as part of the US National Crop Loss Assessment Network Program and the US National Acid Precipitation Assessment Program.

Much of the information that is available on the effects of O_3 on growth and productivity under field conditions has been obtained using open-top exposure chambers. Although widely adopted for gaseous pollutant exposure studies, Runeckles and Wright, and Manning and Krupa, have pointed out that the use of such chambers and the associated experimental protocols for providing a range of O_3 exposure treatments have serious limitations in their abilities to reflect true ambient exposures. Because of significant differences in the microclimate and in the O_3 exposure potential between the chamber and open-field environments, the relevance of the effects observed using such chamber systems to exposures in free air is in question. Nevertheless, the bulk of evidence obtained using a variety of experimental exposure systems clearly indicates the phytotoxic effects of O_3 on crop growth and/or productivity.

As previously noted, any changes in UV-B radiation and surface-level O_3 are only a part of the overall global climate change. In the context of the direct effects of climate change on crops, a key consideration is the increasing concentrations of atmospheric CO_2. While any increases in surface-level UV-B and O_3 can lead to adverse effects on crops over a broad geographic scale, elevated concentrations of $CO_2$are considered to provide a fertilization effect. According to Krupa and Kickert,, an analysis of the available voluminous literature suggests that sorghum, oats, rice, pea, bean, potato, lettuce, cucumber and tomato are among the crop species that appear to exhibit a high degree of responsiveness to the joint effects of all three environmental variables (CO_2, O_3 and UV-B). A critical limitation in this type of assessment is the fact that almost all of our knowledge has been primarily derived from studies on crop response to single rather than to all three variables (CO_2 UV-B or 0). Compounding this limitation is the lack of consideration of the impacts of possible changes in temperature and moisture regimes. Independent of these and other concerns,. As an example, the two most populated regions, People's Republic of China and South Asia, are also the two largest producers of rice and cotton. While rice is considered to be sensitive to elevated levels of UV-B, cotton is known to be sensitive to O_3. Both China and India are regions of high photochemical smog (0) at the present time and most likely will remain so into the future. Because of the virtual lack of data, it is not possible to provide a similar geographic analysis for UV-B at the present time. However, one could expect significant temporal and spatial variability at the local scale for both UV-B and O_3 What is not known with any degree of certainty is how various crops sensitive to elevated levels of UV-B and/or O_3 will respond on a consistent basis to these two variables in the presence of elevated levels of CO_2.

Elevated UV-B Radiation and the Incidence of Crop Pests

According to Runeckles and Krupa, and Manning and v. Tiedemann, the available evidence clearly shows that the effects of UV-B on the incidence and development of pathogen-induced diseases on crops is dependent upon the crop cultivar and age, pathogen inoculum level and the timing and duration of the UV-B exposure. In their studies, Orth *et al.*, exposed three cultivars of cucumber (*Cucumis sativus*) to a daily dose of 11.6 k/m biologically effective UV-B radiation in an unshaded greenhouse before and/ or after inoculation with *Colletotrichum laginarium* (anthracnose) or *Cladosporium cucumerinum* (scab). Pre-inoculation exposure of one to seven days to UV-B resulted in greater disease severity from both pathogens on the susceptible cultivar, Straight-8. Post-inoculation UV-B exposure was much less effective. Although the resistant cultivars Poinsette and Calypso

Hybrid showed increased severity of anthracnose under a heavy pathogen inoculum load when exposed to UV-B (both pre- and post-inoculation), this effect was observed only on the cotyledons and not on the leaves. From their UV-B exposure and crop cultivar studies, Biggs *et al.*, suggested that rust disease on wheat is more likely to exhibit increased severity when a susceptible rather than a resistant cultivar is exposed to UV-B radiation. In contrast, severity of*Cercospora* leaf spot on clonally propagated sugar beet (*Beta vulgaris*) increased in combined exposures to UV-B.

Carns *et al.*, found that when the anthracnose-resistant cucumber cultivar Poinsette was exposed to UV-B doses injurious to the crop, the mycelial growth of *Colletotrichum laginarium* was partially inhibited and spore germination was severely decreased. Owens and Krizek, showed that *Cladosporium cucumerinum* spore germination was also significantly inhibited by UV-B radiation. Similarly, conidial germ tubes of *Diplocarpon rosae* appear to be sensitive to UV-B prior to the penetration of rose leaves. These types of direct adverse effects of UV-B on micro-organisms are well known. In contrast, Orth *et al.*, in their studies on cucumber and *Colletotrichum* concluded that UV-B action on the host was apparently more important than on the fungus *per se,* since there was no difference in the disease severity between plants that received only pre-inoculation UV-B treatment and those that received both pre- and post-inoculation UV-B treatment.

A similar range of types of response is true of the interactions of pathogens and O_3. Effects on the host plant, on the pathogen, or on both may lead to stimulations or inhibitions of disease incidence or severity. Dowding, has stressed the critical importance of the coincidence of the timing of exposure and the infective period to any effect of 0, on the establishment of disease.

As with UV-B, in the case of many fungal pathogens, potential effects of exposure to O_3 at the spore stage appear to be minimal, but the organisms are vulnerable following deposition, since their carbohydrate energy reserves are rapidly depleted on germination. The diverse interactions with O_3 on the leaf surface have been discussed by Dowding, and include effects on cuticular chemistry and surface properties, exuded materials and stomatal response. The growth and development of the pathogen on the host may be inhibited directly by O_3, since toxicity has been observed in axenic culture. Alternatively, important O_3-induced changes in the host may have profound effects on the successful growth and development of the pathogen. Similar interactions may be involved with bacterial pathogens, although infection is usually dependent upon successful entry into host tissues via wounds or insect vectors.

The development of the pathogen may affect the susceptibility of the host plant. Since the early report of 'protection' against 'smog injury' afforded to bean or sunflower leaves by infection with the rust fungi, *Uromyces phaseoli* and *Puccinia helianthi,* respectively, there have been numerous reports of infection with viruses, bacteria and fungi leading to reduced host susceptibility to O_3 . The converse enhancement of the impact of O_3 on host plant growth has been observed with nematode infection. While there is considerable evidence indicating that exposure to O_3 can reduce infection, invasion and sporulation of fungal pathogens, including 'obligate' pathogens such as the rust fungi, examples also exist of increased infection of O_3-injured plants. There is no clear understanding of the mechanisms involved in O_3-host-pathogen-environment interactions, but one generalization that can be made is that pathogens which can benefit from injured host cells and disordered transport mechanisms will be enhanced by earlier exposure of the host to O_3, while those that depend on 'healthy' host tissue will be disadvantaged.

Herbivorous insects and spider mites are major causes of crop loss but little is known of the effects of UV-B on plant-insect interactions and, although there is a sizeable body of information about air pollutant effects, relatively few of these concern O_3. As with pathogens and disease, the topic can be subdivided into the influence of O_3 on insect attack and population dynamics (whether direct or mediated by changes induced in the plant), and the converse effects of insect attack on plant response.

Host plant resistance to insect attack may be modified through metabolic changes which affect feeding preference and insect behaviour, development and fecundity. O_3-induced changes in both major and secondary metabolites may be qualitative and quantitative, and while there is abundant evidence that such changes can influence insect growth and development, there have been few experimental investigations of the specific effects of O_3 Trumble *et al.*, reported that the tomato pinworm (*Keiferia lycopersicella*) developed faster on O_3-injured tomato plants, although fecundity and female longevity were unaffected. the Mexican bean beetle (*Epilachna varivestis*) was found to show a preference for O_3-treated soybean foliage that increased with increased exposure. Such preferences can lead to increased larval growth rates, as shown by the work of Chappelka *et al.*,. Other examples are reviewed in Runeckles and Chevone,.

Only one report appears to exist with regard to insect attack modifying the effects of O_3 on a herbaceous host. Rosen and Runeckles, showed that the combination of extremely low-level O_3 (0.02 ppmv) and infestation with the greenhouse whitefly (*Trialeurodes vaporariorum*) acted synergistically in inducing accelerated chlorosis and senescence of bean leaves. They speculated that the effect might be the result of reaction of O_3 with enhanced ethylene production resulting from whitefly injury.

In summary, the information available on these biotic interactions involving UV-B or O_3 is fragmentary and precludes a clear unravelling of the complexities of the relationships, a situation that will only be remedied by further systematic investigation.

Similarly, as a comparison, the interaction of high CO_2 and plant insect pests has been shown. Lincoln *et al.*, found that insect (butterfly larvae) feeding rates rose as CO_2 in the plant growth atmosphere was increased. This was related to the nitrogen and water content of soybean leaves. In contrast, more recent studies have suggested that leaf-feeding caterpillars do not do as well on plants grown at high CO_2, presumably due to increased carbon: nitrogen ratio (lower nutritive value). These types of studies need to be expanded to determine not only the increases or decreases in the populations of a given insect species, but also in population shifts among multiple species under elevated CO_2 concentrations.

UV-B Radiation and Crop-weed Competition

The dynamics of changes in plant populations and communities is a product of the intra- and inter-specific competition for resources needed for growth and productivity of the competing species, as influenced by abiotic and biotic environmental factors. In view of the ranges in sensitivities of individual species and varieties to both UV-B and O_3, it is to be expected that mixed plantings will show differential responses to either stress. In the case of O_3 stress, a widely held view is that changes in community composition favouring tolerant species will occur.

There have been few investigations of the effects of either UV-B or O_3 on inter-specific competition, and even fewer on the density-dependent intra-specific competition typical of monocultures. With both stresses, plant spacing of monocultures and mixtures will influence the exposures received. In the case of UV-B, this will result from canopy density and shading; with O_3 canopy structure will determine the flux of gas into the foliage as well as dictating the microclimatic conditions. Of the few published studies, some have merely reported differential effects on biomass responses observed with established mixed plantings, e.g., O_3 on grass/clover, while others have attempted to analyse the nature of the competition by the use of replacement series experiments and the calculation of relative crowding coefficients. Although such coefficients provide a measure of competitiveness, Jolliffe *et al.*, have drawn attention to their limitations in interpreting the results of replacement series experiments.

Fox and Caldwell, examined the effects of an artificial increase in UV-B radiation on the competitive interactions of several pairs of species, using replacement series. Three types of associations were included: crop-weed, montane forage species, and weeds of disturbed habitats. The data showed statistically significant shifts in the competitive balance of two of the pairs: *Amaranthus-Medicago* (alfalfa) and*Poa* (bluegrass)-*Geum.* In both cases, UV-B caused a shift in favour of the crop (alfalfa or bluegrass) over the weed species. Only in the case of the *Alyssum-Pisum* (pea) mixture did enhanced UV-B irradiation result in a significant reduction in total mixture biomass, although changes occurred in the proportions of biomass contributed by the individual species of the other pairs. In further studies of the competition between wheat (*Triticum aestivum* cv. Bannock) and wild oat (*Avena fatua*) and between wheat and goatgrass (*Aegilops cylindrica*), W.G. Gold and most recently, Barnes *et al.*, also found a competitive advantage for the crop species (wheat) and increased UV-B enhancement in the former or in both mixtures.

Most of the limited number of studies of the effects of O_3 on competition have utilized crop species. Bennett and Runeckles, used replacement series to study effects on the competition between Italian ryegrass (*Lolium multiflorum*) and crimson clover (*Trifolium incarnatum*). The relative crowding coefficients for ryegrass increased with the O_3 exposure compared to clover on both a dry weight and leaf area basis, indicating a shift in favour of ryegrass.

Other grass-legume studies of the yields of individual competing species have confirmed the effect of O_3 in favouring the grass species. However, recent work by Evans and Ashmore, on a semi-natural grassland dominated by the grasses *Agrostis capillaris, Festuca rubra* and *Poa pratensis* showed that O_3 adversely affected their growth relative to the growth of forb species including clover (*Trifolium repens*) and the major weed species, *Veronica chamaedris, Plantago major*and *Stellaria gramineae.* Since tests of the individual species indicated that, in isolation, the grasses were less sensitive than the forbs, the authors concluded that 'the response of plant communities cannot be predicted from the responses of individual component species'. This contrasts with the conventional wisdom that O_3 will induce changes in community composition favouring tolerant species. While this may well be true in the case of prolonged exposure to high O_3 levels, it appears that the nature of any changes will be highly dependent upon the severity of the O_3 stress imposed on a community and its composition.

There has been no global network for monitoring surface-level UV-B radiation. Long-term UV-B data are sparse and not very reliable. Nevertheless, numerous investigators have examined the effects of UV-B radiation on crops in artificial exposures, but large uncertainties in the relevance to climate change of much of the information obtained remain. According to Runeckles and Krupa,, the transfer of results from growth chamber or greenhouse experiments to the ambient environment has been particularly difficult. This appears to be due to the differences in the characteristics of plants grown under these environments and to photorepair under the high photosynthetic photon flux densities encountered in the ambient environment. Studies of the effects of UV-B (or of 03) on physiological processes such as photosynthesis and on modes of action are appropriately examined under controlled environment conditions. However, the integration of their effects on the processes affected within the whole organism that ultimately lead to growth can only reliably be investigated using plants growing under true field conditions.

Our knowledge of the effects of O_3 is also beset by uncertainties related largely to the lack of information about plant responses under such field conditions. Here the problem is not one of interrelating growth and field observations, but concerns the relevance of the results from the most frequently used open-top exposure chamber method. There is no question of the phytotoxicity of O_3. However, the results obtained in many studies of its effects are primarily supported by controversial statistical techniques and, therefore, the fact remains that the results cannot be validated and show considerable variability from season to season and from location to location most likely because of the types of experimental designs used.

In view of the evidence that suggests that UV-B has little adverse effect on photosynthesis or growth under field conditions, it appears that concern over increases in UV-B irradiation resulting from stratospheric O_3 depletion should focus on longer-term effects probably involving the consequences of damage to nucleic acids. In contrast, increased tropospheric O_3 levels will undoubtedly have immediate adverse effects on most species, independent of any longer-term effects brought about by either adaptation or genetic selection.

In view of the urgency of acquiring information on the potential impacts of the various components of climatic change, including UV-B and tropospheric O_3, that can be realistically envisioned, every effort should therefore be made to avoid wasting research effort and resources on studies that will do nothing to reduce the uncertainties associated with our present information, and which fail to recognize the potential importance of the interactions among the various components.

Combined Effects of Changing CO_2, Temperature, UV-B Radiation and O_3 on Crop Growth

Our understanding of the relationships between crop growth and the atmospheric environment has developed substantially in the past few decades. Improvements in technology have enabled crop physiologists to study processes such as transpiration and carbon dioxide exchange in the field. Mechanistic understanding of the factors controlling photosynthesis and respiration has improved greatly, but we still have no mechanistic understanding of the ways in which many abiotic stresses affect these processes. Our knowledge of factors that control crop development and the partitioning of assimilate between organs is still at the descriptive stage, and poses one of the main limitations on our ability to develop fully mechanistic mathematical models of crop growth. Nevertheless, crop science has reached a stage where reliable semi-empirical simulation models of some of the major world crops have been constructed and used to explore the potential for crop growth and yield in different environments. For the reasons discussed above, these models are restricted in their application to situations where the empirical relationships in them are well defined, but they are often useful in the design of particular experiments to investigate new environments, for example, to apply them to scenarios of climatic change. It should be emphasized that, although we have a much improved understanding of the growth and yield potential of major agricultural crops, this knowledge needs to be extended to include other crops, especially those important for developing countries.

However, as we steadily improve our understanding of how crop yield is influenced by the atmospheric environment, that environment is constantly changing as a result of human activities, and the challenge today is to predict how crops will respond to the changed environments of tomorrow. Traditionally, crop scientists have used controlled environments and field-based facilities to investigate crop responses to individual factors, such as radiation, temperature, humidity and water availability. More recently, manipulation of air quality has been used to study crop responses to air pollutants, such as SO_2, O_3 NO_x, and some of the technology developed for these experiments has been successfully applied to study responses to increased CO_2. Although this 'single component' approach is a highly effective way of discovering how one atmospheric factor influences crops, in practice the atmosphere in the future will alter in several ways simultaneously.

Some of the impending changes are more certain than others. Without doubt, CO_2 concentration is increasing worldwide at about 1.8 mmol/mol per year and under all likely scenarios will continue to do so. This increase will enhance the global greenhouse effect, and the consensus of scientific opinion is that it will result in global warming and in changed global distributions of rainfall and other weather components. The amount of climatic change is, however, very uncertain, because even the best computer simulation models do not adequately describe features such as ocean-atmosphere linkages, and feedbacks associated with changes in atmospheric aerosols or cloud amount and distribution. At present, expert opinion is that, when the CO_2 concentration reaches double its present value of about 350m mol/mol, global temperatures will be about 2.5°C warmer, but this figure and the associated global variability are very uncertain; even more uncertain are any changes in rainfall amount and distribution.

In contrast to the uncertainty associated with computer models, there is clear evidence from satellite and from surface monitoring that stratospheric ozone is being depleted, probably because of emissions of chlorine- and bromine-containing substances, so that globally the column of ozone from the surface through the atmosphere is thinning at about 3% per decade, a rate that seems to be increasing. Because the chemicals believed to be causing this thinning are long-lived in the stratosphere, this thinning process is likely to continue for several decades even though the chemicals are being phased out of production. The consequence of this thinning is that ultraviolet (UV) radiation at the ground, normally severely attenuated by stratospheric O_3 will increase. There is, however, an inadequate network of surface monitoring stations to define the UV climate at the ground (i.e., taking into consideration effects of cloud, dust, etc.) or to determine whether UV has increased significantly. Even if annual mean UV exposure has not changed, there are indications that thinning of stratospheric ozone, similar but less extreme than that observed in the Antarctic, can occur episodically at temperate latitudes.

Air quality in the troposphere is also changing. In particular, in many highly populated regions, emissions of hydrocarbons and exhaust gases from motor vehicles provide the precursors for photochemical production of tropospheric ozone. There is evidence that background tropospheric ozone concentrations have increased by about 10% per decade in Europe over the past 20 years, and that the European background has approximately doubled since the beginning of the century. Although there are no equivalent long-term records, it is likely that the tropospheric O_3, concentration in the United States and Southeast Asia has also increased. Perhaps more important than the mean background concentration of O_3 is the regular occurrence on a regional scale of high concentrations of tropospheric O_3 (typically 5 to 10 times background) associated with stagnant weather systems in summer. These concentrations have been demonstrated to reduce the growth and yield of many agricultural crops.

The atmospheric environment of many world crops of the future will undoubtedly contain more CO_2 than present, is likely to be somewhat warmer, and rainfall (and hence humidity) and cloudiness will probably also change, though in a direction unknown at present. Whether UV will increase much on a seasonal basis remains unclear, but episodes of increased UV at the ground, associated with patches of thinned stratospheric O_3, may occur, especially in temperate latitudes. Without doubt, in regions with large population densities, episodes of high concentrations of tropospheric O_3 and of other pollutants associated with fossil-fuel burning will become more common.

Carbon Dioxide

Several recent reviews of responses of crops to CO_2 have been published, and there is no advantage in repeating the detail of these here. At the leaf level, the two most well-known responses to elevated CO_2 are an increase in the rate of net photosynthesis, P_N, and a decrease in stomatal conductance, g_s. The increase in P is much larger in C_3 species (50 to 100% when CO_2 concentration doubles) than in C_4 species (10%), but a substantial decrease in g (30 to 40%) is observed in both types of species. There is some evidence of photosynthetic acclimation to elevated CO_2, so that increased rates of P_N do not persist at the leaf level in long-term studies with some species, but there is no evidence of acclimation in g. It is not clear whether increased CO_2 also indirectly affects plant growth, e.g., by changing rates of leaf initiation, expansion or longevity, but such effects could have a large influence on crop productivity.

Much of our understanding of effects of CO_2 on plants has been gained from studies with individual leaves. There have been far fewer studies of long-term effects of elevated CO_2 on canopy-scale photosynthesis and transpiration. Most of those reported have been in closed- or open-top chambers, though a few reports of field-scale exposures (FACE) are now becoming available. Drake and Leadley, summarized the data available at that time and concluded that: (1) canopy photosynthesis increases in elevated CO_2 when there is a sink available for the carbon; (2) the relative effect of CO_2 is greatest at highest temperatures; and (3) elevated CO_2 alters many interacting factors, such as canopy architecture and partitioning of assimilates, that mediate gas exchange of canopies and ecosystems.

As a consequence of increased net photosynthesis, dry matter production and yield are substantially increased by elevated CO_2. Several authors have reviewed the literature and concluded that, provided there is adequate water, nutrients and pest control, yields of C_3 and C_4 crops growing in about 700 m mol/mol CO_2 would be about 30 to 40% and 9%, respectively, greater than present yields (350 m mol/mol CO_2) if all other climatic factors were unchanged.

In much of the world, ideal crop growing conditions, with adequate water and nutrition, are wishful thinking, and the influence of elevated CO_2 on aspects such as water-use may be much more relevant. Reviews by Morison, and Eamus, make it clear that, on a leaf area basis, elevated CO_2

improves the water-use efficiency (WUE - the ratio of weight of dry matter produced to weight of water transpired) of plants. On a whole-plant basis, the increase in leaf area resulting from enhanced photosynthesis may cause the amount of water-used per plant to remain close to that of plants growing at current levels of CO_2, although the WUE as defined here increases. We return to this topic later.

CO_2 and Temperature

As discussed earlier, the atmospheric environment of the future is likely to include correlated increases of atmospheric CO_2 concentration and temperature. Long, reviewed the mechanisms by which temperature and CO_2 affect photosynthesis in C_3 species, and developed models of the response of leaf canopy carbon exchange to changes in these variables. He showed that the interaction of carbon dioxide concentration and temperature causes the temperature optimum of the light-saturated rate of CO_2 uptake to increase as CO_2increases. Without this interaction, an increase in global temperature of 2.5°C would increase the frequency of temperatures in temperate regions that are supra-optimal for light-saturated photosynthesis. However, if warming on this scale was coincident with a CO_2concentration increase to above 500 m mol/mol, the temperature optimum would increase, and this inhibition would not occur. Long, went on to develop a simple model of canopy CO_2 uptake (Ac) in response to temperature, taking into account the substantial fraction of the canopy leaf area that would be shaded and therefore not photosynthesizing at the saturated rate.

Interactions between temperature and CO_2 also need to account for effect of temperature on development. When daily temperatures were increased by 3°C and CO_2 was doubled, more rapid development of the crop shortened the growing season, and the potential grain yield (10.4 t/ha) was only 15% larger than in the control, but maturity occurred 30 days earlier. The wheat model used in these simulations contains a number of elements that are sensitive to temperature. Recently, Kocabas, completed a detailed investigation of the model sensitivity to changes in mean temperature at various stages of crop development, and to changes in variability of temperature in various developmental phases and over the whole season. The results indicated the particularly strong dependence of yield on temperature in the phase from emergence to double ridges in wheat, but suggested that changes in temperature variability of ±20% would not have any consistent effects on yield variability. It would be useful to incorporate Long's analysis into the wheat model to allow more precise simulation of the interactive effects of temperature and CO_2 on carbon assimilation and crop development.

Indeterminate crops, such as grass and sugar beet, continue to grow and produce yield as long as the temperature is above a minimum threshold, provided that other factors such as drought and nutrition do not limit growth. Consequently, simultaneous increases in temperature and CO_2 concentration increase the potential yield of such crops in temperate environments, where low temperatures at the beginning and end of the season usually limit productivity.

Perennial tree crops, which yield fruit, nuts and wood, pose complex problems in assessing the effects of CO_2 and temperature interactions, because their yield depends on a phased sequence of development over at least two years. Climatic warming could disrupt this sequence, for example, by providing insufficient winter chilling to synchronize spring bud break, or by advancing ovule development and leading to poor fruit set in spring. Cannell and Smith, pointed out that climatic warming could either advance or delay bud burst in apples depending on the extent to which chilling requirements were met. The long-term influence of CO_2 on tree growth, and its interaction with temperature responses, is very uncertain at present.

CO_2 and Water Availability

The influence of CO_2 on the water-use of crops may well prove to be the most important benefit of increased CO_2 concentrations for agriculture. Morison, reviewed the recent literature and discussed the effect at a range of scales, pointing out that there are now many studies, including annual cereal crops, legumes and woody perennials, which show that, contrary to earlier assumptions, although the absolute amount of dry matter produced decreases at all CO_2 concentrations as water availability is reduced, the *relative* effect of increasing CO_2 on plant growth increases with decreasing water supply. There are at least two mechanisms responsible for this relative enhancement: (1) the reduction in stomatal conductance, with a consequent beneficial effect on leaf water potential and hence on leaf expansion; (2) greater allocation of carbon to roots in elevated CO_2, and hence improved potential for exploring the soil volume for water and nutrients. In contrast, the increased production of leaf area in elevated CO_2 acts to increase the water-use per unit ground area of crops.

It is less clear whether increasing CO_2 will decrease water-use of crop canopies in the field over periods of days or months. This is not only because of the various mechanisms described above, but also because soil evaporation must be accounted for, and feedbacks between leaf temperature, atmospheric humidity and canopy evaporation apply. Consequently, the very rough estimate that a 33% increase in dry matter production combined with a 33% decrease in transpiration in response to doubling CO_2 concentration would increase water-use efficiency by a factor of 2 is unlikely to apply. Goudriaan and Unsworth, used a crop model with feedbacks between vegetation and the atmosphere to estimate that daily plant canopy water-use efficiency would increase by about 35% when carbon dioxide concentration doubled. In a simpler model, but including soil evaporation, Morison, calculated that seasonal water-use by crops with leaf area indices between 2 and 6 would decrease by only 10 to 15% in doubled CO_2 concentration. Nijs *et al.*, grew ryegrass in 350 m mol/mol CO_2 and 600 m mol/mol CO_2. While the total water-use of the canopy was unaffected by growth in elevated CO_2, WUE was increased by 25% on a canopy basis or 87% on a leaf area basis. Results from this experiment indicated that, because the influence of elevated CO_2 on P_N and g_s was greatest early in the season, before the canopy closed, WUE probably varied seasonally, and this may be a general feature for annual crops.

Azam Ali (pers. comm.) used two methods to assess transpiration of crops of sorghum and groundnut growing at 350 and 700 m mol/mol CO_2. Seasonal monitoring of soil water content with a neutron probe, combined with metered irrigation and subsidiary measurements of soil evaporation, enabled cumulative crop transpiration to be calculated. Porometry on individual leaves, and measurements of leaf area distribution, enabled transpiration per unit leaf area to be calculated. In groundnut but not in sorghum, elevated CO_2 reduced transpiration per unit leaf area throughout the season, but more so in the earlier stages

On an even larger scale, de Bruin and Jacob, estimated the influence of doubling CO_2 concentration on regional-scale transpiration, allowing for interactions between transpiration, leaf temperature and the planetary boundary layer. They concluded that, for vegetation where stomatal conductance decreased by 34% in response to elevated CO_2, but leaf area was unchanged, regional transpiration on a typical summer day would decrease by about 11 % for short crops such as cultivated grass and 17% for tall crops such as plantation forests. If negative feedback from the planetary boundary layer had been ignored, these values would have been overestimated (15 and 24%, respectively).

CO_2 and Ozone

When Krupa and Kickert, attempted to review the interactive responses of crops to CO_2 and O_3 they found no publications describing experimental work. We summarize here the few papers that have been published since then, and comment on other work known to be in progress.

Allen, made an estimate of the possible scale by which CO_2 might influence responses of soybeans to O_3. He used published results relating O_3 exposure to yield and a model of the sensitivity of stomatal conductance for soybean to CO_2. According to these models, doubling CO_2 concentration would reduce g_s by about 30% and thus decrease the stomatal uptake of O_3 by the crop. Allen estimated that this would result in a yield decrease of about 15% compared with yield in ambient air. This estimate takes no account of the large direct effects of CO_2 on soybean growth and assumes no interactions between the two gases.

Scientists at the USDA Beltsville Research Center have studied soybeans growing in open-top chambers in the field and exposed to $CO_2$and O_3. Kramer *et al.*,, in a preliminary study, showed that the yield loss attributed to ozone at the site was about 12%, but when CO_2 concentrations were increased to 500 m mol/mol the yield loss was only 6.7%. Mulchi *et al.*, reported a more complex design with three O_3 treatments (charcoal-filtered, CF, non-filtered, NF, NF + 40 m mol/mol O_3) and three CO_2 treatments (ambient, + 50, + 150 mmol/mol). Increasing CO_2 concentration increased grain yield and grain oil content, and decreased protein content. Increasing O_3 reduced grain yield but did not significantly alter oil or protein content. Ozone caused a significant decrease in yield compared to the charcoal-filtered controls. When CO_2 was added to the ozone treatments, it partially counteracted the negative effect of O_3 so that yields in chambers with plus 150 m mol/mol CO_2 and NF air were similar to those in ambient CO_2 and CF air. The results suggested that most effects of CO_2 and O_3 were additive.

Barnes and Pfirrman, grew radishes in a phytotron in Munich, Germany, at two CO_2 concentrations (385 and 765 m mol/mol) and two O_3 concentrations (20 nmol/mol and 73 nmol/mol), and studied gas exchange, growth and productivity, and mineral composition of the crop. In clean air the typical effects of elevated CO_2 on photosynthesis (increased) and stomatal conductance (decreased) were observed; the extra carbon assimilation stimulated root growth rate by 43% but there was no significant effect on shoot growth or leaf area. At ambient CO_2, the high O_3 treatment depressed photosynthesis by 26% (compared to the low treatment) and induced a slight reduction in g_s, with the net result that water-use efficiency declined. The reduction in carbon uptake was reflected in reduced growth, with roots more affected than shoots. Interactive effects of ozone and CO_2 were generally consistent with a reduction in ozone-induced responses in elevated CO_2. Early in the experiment, elevated CO_2 reduced the effect of O_3 on photosynthesis, but later this benefit disappeared, perhaps because long-term exposure to O_3 has an overall limiting effect on plant response to CO_2. The effects of CO_2 and O_3 on g_s appeared to be additive. Work is in progress in Munich on responses of grass swards to O_3, CO_2 and water availability. Preliminary results indicate that the growth depression which is observed as water is withheld in ambient CO_2 is alleviated at higher CO_2 levels. Responses of yield and water-use efficiency to the mixtures of CO_2 and ozone seem to be additive rather than synergistic.

In Newcastle, England, work is in progress exposing winter and spring cultivars of wheat to two CO_2 concentrations and two O_3concentrations. Early results indicate that after 50 days' exposure there was evidence of downregulation of photosynthesis in response to prolonged exposure to elevated CO_2; this was most pronounced in the winter cultivars. Long-term exposure to O_3 resulted in a decrease in the light-saturated rate of CO_2 assimilation, partial stomatal closure, and accumulation of water soluble carbohydrate and starch in leaves. These resulted in decreased growth, with root growth more severely affected than shoot growth. In plants exposed to elevated CO_2 and O_3, effects on stomatal conductance were less than additive, and CO_2 enhancement effects on photosynthesis and growth were reduced by O_3.

In summary, from the very limited experimental results available at the time of writing, interactions between exposures to elevated O_3 and CO_2 seem to result in approximately additive effects on growth and yield. Both gases decrease stomatal conductance, and this results in less uptake of O_3 in elevated CO_2 treatments than in ambient air. Several authors have noted that this reduces (or at least delays) the damaging effects of O_3 In principle, the reverse is also true, i.e., increased O_3, by reducing stomatal conductance, negating some of the benefits to growth that elevated CO_2 confers. However, the O_3 exposure necessary to induce stomatal closure is generally sufficient to cause visible leaf injury, and this disbenefit is likely to be far more damaging than reduced CO_2 uptake through stomata. More field-based studies are necessary to assess fully the effects of the two gases in combination.

CO_2 and UV-B Radiation

Many studies of effects of UV-B radiation on crops are seriously flawed because UV exposures were either inadequately specified and/or were unrealistically large. Tevini,, in a thorough review of effects of UV-B radiation on plants, pointed out that, since the spectrum of UV from artificial light sources differs from that in the solar spectrum, and since many photobiological processes are strongly wavelength dependent, it is important to use weighting functions based on the action spectrum for specific responses before experiments in different exposure systems can be compared. Unfortunately, action spectra for many responses remain unknown. In addition, many plant responses to UV-B are larger when plants are growing at low light levels typical of growth chambers and some glasshouses than they are in full sunshine, probably because natural protective pigments are inadequately synthesized and repair processes are restricted in low light. For this reason, field-based studies of crops are most relevant for estimating responses of productivity and yield to UV-B. Even so, most field studies have used continuous UV-B exposures corresponding to 15 to 25% reductions in stratospheric O_3 i.e. roughly 30 to 50% increases in UV-B above present values. Such increases are much larger than are likely to occur as long-term means in agricultural regions under any likely depletion of stratospheric ozone in the next 30 years.

Effects of UV-B on yield of crops growing in the field have produced very variable results. Work in the late 1970s and early 1980s by Biggs and colleagues in Florida, using very high UV exposures, found yield reductions in only about half the crops studied. Work in Germany, also at high UV exposures, found no UV-B effects on cabbage, lettuce and rape. More recently, a large field-based study in Beltsville, Maryland, by Teramura, has revealed a high degree of intra-specific variability among soybean cultivars. The two most sensitive cultivars were grown for five seasons in the field in two UV treatments, corresponding to 16 and 25% ozone depletion simulations. In one cultivar (Essex) yield was reduced by 20 to 25% at the high UV treatment; in the other cultivar, yield generally increased by 10 to 22% in this treatment. Neither cultivar showed a consistent significant yield change at the lower UV exposure. The conclusion from these studies must be that, under any realistic scenario of increasing UV-B radiation, crop yields are unlikely to be altered to a significant extent. We therefore turn to the question of whether UV-B exposure alters the sensitivity of crops to other atmospheric changes.

Effects of the combination of elevated CO_2 and UV-B radiation on crop growth and yield have been addressed in very few studies. Teramura and his colleagues grew wheat, rice and soybean in a glasshouse experiment. Treatments in the factorial design were: ambient CO_2 (350 (m mol/mol), ambient UV, elevated CO_2 (650 m mol/mol) and elevated UV-B (corresponding to a 10% decrease in stratospheric O_3 at the equator). Compared to the control, seed yield and total biomass increased significantly in elevated CO_2 for all three species. However, when UV-B and CO_2 were increased simultaneously, no increase in either seed yield (wheat and rice) or total biomass (rice) was observed compared to the control. In contrast, in soybean, the increases in seed yield and biomass induced by CO_2 alone were maintained in the high CO_2/high UV-B environment. Studies of leaf gas exchange indicated that UV-B reduced the apparent carboxylation efficiency in wheat and rice, but not in soybean.

Results of a CO_2/UV study on seedlings of sunflower and maize reported by Tevini, are confounded by temperature changes in some treatments. A 25% increase in UV-B at 340 m mol/mol CO_2 changed sunflower and maize dry weights (W) by -14 and -24%, respectively (compared to ambient controls). The same UV-B increment combined with a +2°C temperature change altered W by +5 % in sunflower and +31 % in maize. Adding a doubled CO_2 concentration to the elevated UV and temperature regime altered W by +19 and +32% for sunflower and maize, respectively. The incomplete experimental design limits interpretation of this study, but the results suggest that temperature and CO_2 are much more important influences on growth than any likely realistic increases in UV-B.

Ozone

There are a number of recent reviews of the mechanism of action of O_3 on plants, and of crop responses to O_3; consequently, we will only briefly summarize the main features. The response of plants to O_3 may be viewed as the culmination of a sequence of physical, biochemical and physiological events. The O_3 diffuses from the air into the leaf through stomata, and these exert control on O_3 uptake. Plants are able to detoxify O_3 or its metabolites and can repair or compensate for O_3 impacts, so visible signs of ozone injury may not occur if the rate of O_3 uptake is sufficiently small. When resources have to be devoted to repair or compensation mechanisms, or when O_3 uptake is too large for full repair to be achieved, growth and yield may be reduced in the absence of visible injury by O_3. The principal modes of action of O_3 on plants are through injury to proteins and membranes, reduction in photosynthesis (i.e., carbon gained), changes in allocation of carbohydrate, and acceleration of senescence. The main result of O_3exposure is therefore a reduction in the capacity of the plant to accumulate photosynthate through loss in photosynthetic capacity and increased foliage senescence.

Methods for exposing crops to ozone in experiments designed to develop yield/exposure responses are well established. The most commonly used technique is the open-top field chamber, which enables crop stands to be grown in either charcoal-filtered air or in air to which known amounts of O_3 have been added. Coordinated programmes in the United States and Europe have investigated crop responses to O_3 in open-top chambers, and relationships between yield and seasonal mean O_3 concentration are available for many of the major world crops. The seasonal mean index, however, has been shown to be inadequate in relating ozone exposure to effects, primarily because of its inability to consider duration of exposure and its consideration of all concentrations as equal in their effect. Both duration and concentration are important in the effect of ozone. In particular, studies have demonstrated that ozone's effect is cumulative and that higher concentrations are more important than lower concentrations in causing an effect. Consequently, the cumulative, peak-weighted indices are more appropriate because of their biological relevancy. Examples of such indices include SUM06 (in units of nmol/mol-h) which cumulates all concentrations equal to or greater than 0.06 nmol/mol, or SIGMOID which cumulates all concentrations during the season and weights all concentrations according to a sigmoid function. The US Environmental Protection Agency (EPA) Draft Ozone Criteria

Document, reviewed and tabulated the latest information concerning effects of ozone on crop yields using both the seasonal mean, and two of the cumulative, peak-weighted indices. The data for crop yield response are primarily from the National Crop Loss Assessment Network (NCLAN). Yield reductions in the NCLAN studies are calculated relative to a typical background O_3 concentration of SUM06 of O_3 or a 7-h seasonal mean of 25 nmol/mol. Yield losses of 10% or less in 50% of the crops (12 crop species, 54 studies including well-watered and drought studies) would occur with SUM06 exposure concentrations of 26.4 nmol/mol-h in three months, or 49 nmol/mol 7-h seasonal mean. Some species/cultivars were particularly sensitive; 11 % would be expected to have a yield reduction of 10% at a 7-hr seasonal mean of less than 35 nmol/mol. Similarly, 18% of the crops/ cultivars are expected to have yield reductions of 10% or less at 3-month SUM06 concentrations of less than 10 nmol/mol-h. Both of these exposure values are quite low compared to the decade-average (1982-1991) exposure (3 months) from all monitoring sites across the United States (SUM06 = 29.5 nmol/mol-h; 7-h mean = 54 nmol/mol). In general, grain crops are less sensitive than others, but within-species variability and sensitivity may be greater than between species. Krupa and Kickert, reviewed various published lists of the sensitivity of crop yields to O_3 onion, spinach, potato, alfalfa and cotton appear to be particularly sensitive crops, although it is possible to find resistant cultivars of all these species. Legumes range between sensitive and intermediate, and cereal and grass crops tend to be relatively resistant.

Ozone and Temperature

When interactions between ozone exposure and temperature were reviewed by the US EPA, it was concluded that there were variable and conflicting results. More recent analysis suggests that many of these studies in controlled environments confounded effects of temperature with changes in vapour pressure deficit (VPD) because relative humidity was kept constant while temperature was increased. Since VPD can have profound effects on stomatal responses, evaporation rates and leaf expansion, it is not surprising that these early results have led to confusion. Recently, Todd *et al.*, designed an experiment with tomato seedlings in which differences in VPD at different temperatures were minimized during O_3 exposure. This study showed that, out of 11 growth variables measured, the only significant modifications by temperature of the effects of O_3 were on stem fresh weight and specific leaf area (leaf area/leaf dry weight). The authors suggested that VPD probably plays a more important role in determining sensitivity to O_3 than temperature. It is important to have more studies to clarify this issue, as many field-based studies of ozone responses use open-top chambers in which temperatures are increased by a few degrees above ambient. If either the changed temperature or the changed VPD in such chambers alters the sensitivity of the crops to O_3 there would be serious consequences for the application of much of our knowledge of the O_3 sensitivity of crops that has been gained from field-chamber experiments.

A second important O_3/temperature interaction, particularly affecting perennial species, including trees, is winter hardiness. Several studies have shown that exposures to O_3 reduce the frost-hardiness; Davison *et al.*, reviewed the subject. Whilst most of this research has been with woody species, Barnes *et al.*, showed that daily exposures to about 80 nmol/mol O_3 for 7 days significantly reduced the survival of pea seedlings after exposure to nighttime temperatures of about -4°C. It seems likely that, as with sulphur dioxide, exposure to ozone could alter the frost hardiness of a number of agricultural crops. This response would be most likely to be significant in the event of an early autumn frost on early-sown crops such as winter cereals. There are suggestions that exposure of trees in summer to ozone alters the hardiness the following winter. Eamus and Murray, pointed out that, even in severe winters, there are brief periods of mild temperatures that induce partial dehardening of woody species. They speculated that O_3 increases tree predisposition to dehardening in such conditions and consequently puts trees at greater risk from low temperatures. There are likely to be significant differences between species in this phenomenon. For example, in Florida, Eissenstat *et al.*, found that, although O_3 reduced frost hardiness of citrus and avocado, the effects were small and the likelihood of significantly changing frost resistance is slight.

Ozone and Water Availability

The availability of soil water is one of the strongest influences on crop growth and productivity. When available, irrigation is commonly used to alleviate water stress, but in conditions of large evaporative demand, even irrigated crops may experience water stress. The most immediate effects of water stress are on plant-water potential, and may lead to reductions in leaf expansion rates and altered partitioning of dry matter, with increased assimilate partitioning to roots. When stomatal closure occurs, there is a resulting reduction in transpiration and CO_2 uptake and an increase in leaf temperature.

Water Stress and O_3 Responses

There has long been a general belief that water stress reduces the magnitude of any adverse effects of O_3 i.e., leaf injury, growth and yield reductions. This belief was based on observations that stomatal closure, induced by drought, reduced the rate and quantity of O_3 absorbed by leaves. However, more recent results show that the O_3/water availability interaction is complex, and probably depends on the magnitude and timing of the water stress (which has seldom been adequately defined in air pollution experiments). One difficulty in quantifying the interaction is the very large effect that water stress alone may have on yield, so that any influence of O_3 is sometimes masked.

Temple *et al.,* studied the effects of O_3 and water stress on cotton growing in California. In this typically hot, dry season, the water-stressed plants wilted frequently, and yielded much less than the well-watered controls; nevertheless, when compared to yields of plants in a background O_3 concentration of 25 nmol/mol, and expressed as a percentage, the estimated yield loss at a seasonal mean 0, concentration of 50 nmol/mol was 7% in the well-watered plants and 2% in the water-stressed crop. Clearly the absolute yield loss from drought was much larger than the yield loss attributed to O_3 in either treatment.

Heagle *et al.*, reviewed six studies of soybean responses to O_3 and water stress, and concluded that in only three were there significant interactions, i.e., the clear negative relationships between yield and O_3 exposure observed with well-watered plants were much reduced with water stressed plants. More recently, Heggestad and Lesser, analysed three years of data for four soybean cultivars. They concluded that, in most cases, the relationships between yield and O_3 concentration had similar slopes for the water-stressed and well-watered treatments. If this conclusion applies more generally, the extra yield losses which would result from O_3 exposure could be estimated if a yield response to drought alone is known. Clearly, there is still uncertainty over the influence of water stress on the form of the yield response to O_3. This uncertainty is unlikely to be resolved until better experimental designs with more precise specifications of the degree of water stress are developed.

O_3 and Water-Use Efficiency

For irrigated crops where water supplies may be limited or expensive, it is important to know whether O_3 would influence the water-use efficiency (WUE) and consequently the amount of water required in a season. Reich *et al.*, exposed well-watered soybeans to 130 nmol/mol O_3 for 7 hours daily, and found a 25% decrease in WUE compared with controls in 10 nmol/mol O_3. Similar results were found for alfalfa, probably indicating that carbon dioxide uptake was reduced more by O_3 than water loss. Tingey *et al.*, also reported a significant decrease in WUE in soybeans exposed to episodic exposure regimes of 20-30 nmol/mol-h (3-month SUM06). The study found the O_3 did not close stomata, but rather affected CO_2 assimilation. A significant increase in leaf construction cost was reported, indicating that both carbon and water are used inefficiently by plants exposed to ozone. In contrast, Greitner and Winner, found that O_3 exposure *increased* the WUE of radish and soybeans. This disagreement probably reflects variation with exposure dynamics, stage of development, and genetic variation between and within species.

Recently Barnes and Pfirrman, investigated effects of O_3 and elevated CO_2 singly and in combination on the WUE of radish. Both gases reduced stomatal conductance, and the combination of gases reduced conductance still further.

In summary, relatively severe water stress may reduce the yield losses attributable to O_3 in some crop species but, in general, the yield loss resulting from the water stress outweighs the benefits of the O_3 protection. For well-watered crops, O_3 exposure may change the water-use efficiency compared with crops in clean air, but the direction and magnitude of change probably depend on exposure and genetic factors. The influence of O_3 and CO_2 in combination on WUE is potentially important and merits further study.

Ozone and UV-B radiation

Krupa and Kickert, found no reports of O_3/UV-B interactions in their review. They assessed potential risks on a geographical basis, using distributions of major crops, tropospheric ozone concentrations and UV-B irradiance. They suggested that interactions might involve episodic exposure to the two stresses, so that peaks of O_3 would coincide with lower UV-B irradiance and vice versa. Runeckles and Krupa, have developed this concept further, arguing that, as tropospheric O_3 increases, UV-B irradiance at the surface is reduced, because the O_3 absorbs some UV. In theory, this effect could be important, in spite of the relatively small contribution of tropospheric O_3 to the total atmospheric O_3 column, because scattering by aerosols and molecules in the troposphere increases the radiation path length. In an attempt to quantify this effect, Albar, compared the measured solar spectral irradiance I (1) at the ground near Nottingham, England, on two days (12 and 18 July 1990) when the tropospheric O_3 concentration was 51 and 84 nmol/mol, respectively, and stratospheric O_3 was constant.

As mentioned earlier, many investigations of plant responses to UV-B are seriously flawed by inadequate specification and/or excessive UV exposure. An improved field-based system at Raleigh, North Carolina, using open-top chambers for studying crop responses to UV and O_3 has been used for three seasons of research on soybeans. The system exposed soybean crops from emergence to maturity to UV treatments (ranging from ambient to about twice ambient biologically effective UV-B) and to O_3treatments giving seasonal mean 12 h/d O_3 concentrations from 14 to 83 nmol/mol.

The ozone treatments resulted in reductions of photosynthesis, accelerated senescence, and reduced yield, in agreement with many other published studies. In contrast, the UV treatments, even at this relatively large UV irradiance, did not induce any significant changes in photosynthesis or yield, and there were no UV/O_3 interactions.

Fiscus *et al.*, attempted to reconcile the lack of response to UV reported in the Raleigh experiments with other reports which indicate that increased UV-B causes physiological dysfunction and reductions in crop yields. They concluded that three factors make the conclusions of other field and glasshouse studies, at best, hard to interpret and possibly misleading: a frequent failure to monitor UV-B adequately; a tendency to underestimate UV exposures when relying on model calculations which do not allow sufficiently for effects of dust in the atmosphere; and no adjustment of UV exposure for seasonal and daily weather changes (and hence a tendency towards unrealistically large exposures).

References

Hall, A.E. 1993. Breeding for heat tolerance. *Plant Breed Res.* 10: 129-168.

Howarth, C.J. 1991. Molecular response of plants to an increased incidence of heat shock. *Plant Cell Environ.* 14: 831-841.

Idso, S.B., Kimball, B.A., Anderson, M.G. and Mouney, J.R. 1987. Effects of atmospheric CO_2 enrichment on plant growth; the interactive role of air temperature. *Agric. Ecosystems & Environment* 20: 1-10.

Kunst, L., Browse, J. and Somerville. 1989. Enhanced thermal tolerance in a mutant of *Arabidopsis* deficient in palmitic acid unsaturation.*Plant Physiol.* 91: 401-408.

Thomas, R.B., Reid, C.D., Ybema, R. and Strain, B.R. 1993. Growth and maintenance components of leaf respiration of cotton grown in elevated carbon dioxide partial pressure. *Plant Cell Environ.* 16: 533-546.

Ziska, L.M. and Bunce, J.A. 1993. Inhibition of whole plant respiration by elevated CO_2 as modified by growth temperature. *Physiol. Plant.*87: 459-466.

8

Potential Effects of Climate Change on Food Production and Security

Since the late 1950s, global agricultural output has increased at rates and to levels that are unprecedented in human history. Much of the productivity increase is attributed to the breeding of high-yielding crop varieties, intensive use of inorganic fertilizers and pesticides, expansion of irrigation, and capital-intensive farm management.

In the 1970s, the euphoria surrounding the 'Green Revolution' was questioned in the wake of the energy crisis and growing awareness of long-term environmental consequences. Concern over soil erosion, groundwater contamination, soil compaction and decline of natural soil fertility, and destruction of traditional social systems, led to a reappraisal of what were then considered to be the most advanced agricultural production techniques. Since then, agricultural research has expanded its scope to include sustainable and resource-efficient cropping systems and farm management practices.

Since the beginning of the 1980s yet another threat to agriculture has attracted much attention. Many climatologists predict significant global warming in the coming decades due to increasing atmospheric carbon dioxide and other trace gases. As a consequence, major changes in hydrological regimes have also been forecast to occur. The magnitude and geographical distribution of such climate-induced changes may affect our ability to expand food production as required to feed a population of more than 10 000 million people projected for the middle of the next century. Climate change could have far-reaching effects on patterns of trade among nations, development, and food security.

Beyond what is known about greenhouse gases and the climate system, however, lie great uncertainties: How much warming will occur, at what rate, and according to what geographical and seasonal pattern? What secondary processes will the warming trend induce, and what might be the physical and biological impacts of such processes? Will some areas benefit while other areas suffer, and who might the winners and losers be? And, if such damages are unavoidable, what can be done to adapt or modify our systems so as to minimize or overcome them? These are important and complex questions, and we have only begun to understand them and to develop methods for their analysis.

Recent research has focused on regional and national assessments of the potential effects of climate change on agriculture. For the most part this work has treated each region or nation in isolation, without relation to changes in production elsewhere and without paying attention to climate change effects on the world market and their feedbacks. Assessments of potential impacts have been achieved in national studies completed in the United States, Australia, and the United Kingdom. Regional studies have been conducted in high-latitude and semi-arid agricultural areas. These regional and national studies have been summarized in the IPCC Working Group II Reports.

In 1989 the US Environmental Protection Agency (EPA), with additional support provided by the US Agency for International Development (USAID), commissioned a three-year study on the effects of climate change on world food supply. The present study is an initial attempt to arrive at an integrated global assessment of the potential effects of climate change on agriculture and the world food system. The collaborative project was jointly managed by the Goddard Institute of Space Studies (GISS) and the Environmental Change Unit (ECU) in collaboration with the International Institute for Applied Systems Analysis (IIASA) and involved about 50 scientists worldwide.

Climate Change Scenarios

Scenarios of climate change were developed in order to estimate their effects on crop yields and food trade. A climate change scenario is defined as a physically consistent set of changes in meteorological variables, based on generally accepted projections of CO_2 (and other trace gases) levels. The range of scenarios analysed is intended to capture the range of possible effects and to set limits on the associated uncertainty. One set of scenarios for this study was created by changing observed data on current climate according to the results of doubled CO_2 simulations of three General Circulation Models (GCMs).

The temperature changes of these GCM scenarios (4.0-5.2°C) are at or near the upper end of the range (1.5-4.5°C) projected for doubled CO_2 warming by the IPCC. The GISS and GFDL scenarios, however, are near the mean temperature change (3.8°C) of recent doubled CO_2 experiments documented for atmospheric GCMs with a seasonal cycle and a mixed-layer ocean.

GCMs currently provide the most advanced means of predicting the potential future climatic consequences of increasing radiatively active trace gases. They have been shown to simulate current temperatures reasonably well, but do not reproduce current precipitation accurately; and their ability to reproduce current climate varies considerably from region to region. Of special importance for agricultural climate change impacts, there is a notable lack of consensus among GCMs in prediction of regional soil moisture changes. Furthermore, GCMs have so far not been able to produce reliable projections of changes in climate variability, such as alterations in the frequencies of drought and storms, even though these could significantly affect crop yields.

For the crop modelling part of this study, climate changes from doubled CO_2 GCM simulations are utilized with an associated level of 555 ppm CO_2; the assumed timing of the simulations with the world food model is that these conditions will occur in year 2060. Rates of future emissions of trace gases and the point in time when their effects will be fully realized are not certain. Because other greenhouse gases besides CO_2 such as methane (CH_4), nitrous oxide (N_2O), and the chlorofluorocarbons (CFCs), are also increasing, an 'effective CO_2doubling' has been defined as the combined radiative forcing of all greenhouse gases having the same forcing as doubled CO_2 (usually defined as ~600 ppm). Level of CO_2 is important when estimating potential impacts on crops, because crop growth and water use have been shown to benefit from increased levels of CO_2. A CO_2 level of 555 ppm was associated with the effective doubled CO_2 climate projections for use in the crop modelling simulations. This was based on the GISS GCM transient trace gas scenario A described in Hansen *et al.*,, in which the simulated climate had warmed to the effective doubled CO_2 level of about 4°C by 2060. This level assumes that non-CO_2 trace gases contribute ~15% of the change in radiative forcing from 300 to ~600 ppm.

Two other climate change scenarios were tested. The GISS transient scenario consists of a separate GCM run with gradually increasing atmospheric CO_2 levels. The CO_2 concentrations in the GISS transient scenario A were assumed to be 405 ppm, 460 ppm and 530 ppm, respectively, in the decades of the 2010s, 2030s and 2050s. Crop modelling experiments were conducted separately for these three time periods.

Another scenario, termed GISS-A, utilized the climate changes projected for the 2030s from the GISS transient run with 555 ppm CO_2 for the crop model simulations. This scenario was used to test the consequences of a lower sensitivity of the climate system to increasing atmospheric greenhouse gas concentrations. The GISS-A scenario projects a global temperature rise of 2.4°C and a 5% increase in precipitation.

Estimation of Site-Specific Potential Changes in Crop Yields

Crop models and a decision support system developed by the International Benchmark Sites Network for Agrotechnology Transfer were used to estimate how climate change and increasing levels of carbon dioxide may alter yields of major crops. The simulations represented both major production areas and vulnerable regions at low, mid and high latitudes. The IBSNAT models simulate crop growth and yield formation as influenced by genetics, climate, soils and management practices. Models used were for wheat, maize, rice and soybean.

The crop models account for the beneficial physiological effects of increased CO_2 concentrations on crop growth and water use. Most plants growing in experimental environments with increased levels of atmospheric CO_2 exhibit increased rates of net photosynthesis and reduced stomatal openings, thereby reducing transpiration per unit leaf area while enhancing photosynthesis. Crop modelling simulation experiments were conducted for baseline climate (1951-1980) and GCM doubled CO_2 climate change scenarios with and without the physiological effects of CO_2.

The study also tested the efficacy of farm-level adaptations to climate change, including change in planting date, change of cultivar, irrigation, fertilizer, and change of crop. These measures were then grouped into two levels of adaptation: Level 1 implies little change to existing agricultural systems reflecting relatively easy and low-cost farmer response to a changing climate. Level 2 implies more substantial changes to agricultural systems possibly requiring resources beyond the farmer's means. It must be noted that costs of adaptation and future water availability for irrigation under the climate change scenarios were not considered in the study.

Aggregation of Crop Modelling Results

Data on crop yield changes expected for different scenarios of climate change had to be compiled for all crop commodities and geographical groupings represented in the IIASA/FAP world food model, the Basic Linked System (BLS).

Crop model results for wheat, rice, maize and soybean from the selected sites were aggregated by weighting regional yield changes, based on current production, to estimate changes in national yields. The regional yield estimates represent the current mix of rainfed and irrigated production, the current crop varieties, nitrogen management and soils. Production data were gathered by scientists participating in the study and from the FAO, the USDA Crop Production Statistical Division, and the USDA International Service.

Changes in national yields of other crops and commodity groups and of regions not simulated with crop models were estimated based on similarities to modelled crops and growing conditions, and previous published and unpublished climate change impact studies. Estimates were made of yield changes for the three GCM scenarios with and without direct effects of CO_2. The yield changes with the direct effects of CO_2 were based on the mean responses to CO_2 for the different crops in the crop model simulations.

In total, 12 climate change yield impact scenarios were developed: for each of three climate models (GISS, GFDL, UKMO) four scenarios are specified (GCM without direct effects of CO_2 on yields; with direct effects of CO_2 on yields; with direct effects of CO_2 and Adaptation Level 1; with direct effects of CO_2 and Adaptation Level 2).

Simulation of Climate Change Yield Impacts on the World Food SystemThe agricultural production components of the national models in the BLS were modified to incorporate the projected changes in yields and crop productivity. Impacts were assessed for the period 1990 to 2060, with population growth and technology trends projected to that year. The simulation results obtained under each of the climate impact scenarios are compared to a 'neutral' point of departure, a BLS reference scenario which assumes that policies, especially those affecting economic and technological development remain more or less unchanged. The reference scenario projects a possible future based on an internally consistent set of assumptions. The difference between the reference scenario and the climate change simulations is the climate-induced *dynamic* effect. Note that the results of this study are considered to be long-term projections, not forecasts. The considerable length of the projection period makes it impossible to avoid judgement concerning some of the assumptions in the system.

Crop Models and Yield Simulations

The IBSNAT crop models were used to estimate how climate change and increasing levels of carbon dioxide may alter yields of world crops at 112 sites in 18 countries. The crop models used were CERES-Wheat, CERES-Maize, CERES-Rice and SOYGRO.

The IBSNAT models are comprised of parameterizations of important physiological processes responsible for plant growth and development, evapotranspiration, and partitioning of photosynthate to produce economic yield. The simplified functions enable prediction of growth of crops as influenced by the major factors that affect yields, i.e., genetics, climate (daily solar radiation, maximum and minimum temperatures, and precipitation), soils, and management practices. The models include a soil moisture balance submodel so that they can be used to predict both rainfed and irrigated crop yields. The cereal models simulate the effects of nitrogen fertilizer on crop growth, and these were studied in several countries in the context of climatic change. For the most part, however, the results of this study assume optimum nutrient levels.

The IBSNAT models were selected for use in this study because they have been validated over a wide range of environments and are not specific to any particular location or soil type. The validation of the crop models over different environments also improves their ability to estimate effects of changes in climate. Furthermore, because management practices, such as the choice of varieties, planting date, fertilizer application and irrigation, may be varied in the models, they permit experiments that simulate adaptation by farmers to climate change.

Physiological Effects of CO_2

Ratios were calculated between measured daily photosynthesis and evapotranspiration rates for a canopy exposed to high CO_2 values, based on published results, and the ratios were applied to the appropriate variable in the crop models on a daily basis. The photosynthesis ratios (555 ppm CO_2/ 330 ppm CO_2) for soybean, wheat, rice and maize were 1.21, 1.17, 1.17 and 1.06, respectively. Changes in stomatal resistance were set at 49.7/34.4 s/m for C_3 crops and at 87.4/55.8 s/m for C_4 crops, based on experimental results by Rogers *et al.*,. As simulated in this study, the direct effects of CO_2 may bias yield changes in a positive direction, since there is uncertainty regarding whether experimental results will be observed in the open field under conditions likely to be operative when farmers are managing crops. Plants growing in experimental settings are often subject to fewer environmental stresses and less competition from weeds and pests than are likely to be encountered in farmers' fields. Recent field free-air release studies have found overall positive CO_2 effects on cotton under current climate conditions.

Crop Yield Simulations

Crop modelling simulation experiments were performed for baseline climate (1951-1980), and GCM doubled CO_2 climate change scenarios with and without the physiological effects of CO_2. This involved the following tasks:

1. For the countries studied, geographical boundaries were defined for the major crop production regions; agricultural systems (e.g., rainfed and/ or irrigated production, number of crops grown per year) were described, and data on regional and national rainfed and irrigated production of major crops were gathered.
2. Observed climate data for representative sites within these regions were obtained for the baseline period (1951-1980), or for as many years of daily data as were available; the soil, crop variety, and management inputs necessary to run the crop models at the selected sites were specified.
3. The crop models were validated with experimental data from field trials, to the extent possible.
4. The crop models were run with baseline data, and GCM climate change scenarios, with and without direct effects of CO_2 on crop growth.

Rainfed and/or irrigated simulations were carried out as appropriate to current growing practices.

5. Alterations in farm-level agricultural practices that would lessen any adverse consequences of climate change were identified and evaluated, by simulating irrigated production and other adaptation responses, e.g., shifts in planting date and substitution of crop varieties.

Farm-Level Adaptations

In each country, the agricultural scientists used the crop models to test the possible responses to the worst climate change scenario (this was usually, but not always, the UKMO scenario). These adaptations included change in planting date, change of cultivar, irrigation, fertilizer and change of crop. Irrigation simulations in the crop models assumed automatic irrigation to field capacity when plant available water dropped to 50%, and an irrigation efficiency of 100%. Not all adaptation possibilities were simulated at every site and country: choice of adaptations to be tested was made by the participating scientists, based on their knowledge of current agricultural systems.

The adaptation simulations were not comprehensive because not all possible combinations of farmer responses were tested at every site. Spatial analyses of crop, climatic and soil resources are needed to test fully the possibilities for crop substitution. Neither the availability of water supplies for irrigation nor the costs of adaptation were considered in this study; these are both critical needs for further research.

Effects on Crop Yields

Depending on present conditions, global warming and CO_2 enrichment can have positive or negative impacts. Simulated yield increases in the mid and high latitudes are caused primarily by:

1. *Positive physiological effects of' CO_2* At sites with cooler initial temperature regimes, increased photosynthesis more than compensated for the shortening of the growing period caused by warming.

2. *Lengthened growing season and amelioration of cold temperature effects on growth.* At some sites near the high-latitude boundaries of current agricultural production, increased temperatures extended the frost-free growing season and provided regimes more conducive to greater crop productivity.

3. The primary causes of decreases in simulated yields are:
4. *Shortening of the growing period.* Higher temperatures during the growing season speed annual crops through their development (especially the grain-filling stage), causing less grain to be produced. This occurred at all sites except those with the coolest growing-season temperatures in Canada and the former USSR.
5. *Decrease in water availability.* This is due to a combination of increases in evapotranspiration rates in the warmer climate, enhanced losses of soil moisture and, in some cases, a projected decrease in precipitation in the climate change scenarios.
6. *Poor vernalization.* Vernalization is the requirement of some temperate cereal crops, e.g., winter wheat, for a pcriod of low winter temperatures to initiate or accelerate the flowering process. Low vernalization results in low flower bud initiation and ultimately reduced yields. Decreases in winter wheat yields at some sites in Canada and the former USSR were caused by lack of vernalization.

Crop Yields Without Adaptation

The magnitudes of the estimated yield changes vary by crop. Global wheat yield changes weighted by national production are positive with the direct CO_2 effects, while maize yield is most negatively affected, reflecting its greater production in low-latitude areas where simulated yield decreases are greater. Maize production declines most with direct CO_2 effects, probably due to its lower response to the physiological effects of CO_2 on crop growth. Simulated soybean yields are most reduced without the direct effects of CO_2, but are least affected in the less severe GISS and GFDL climate change scenarios when direct CO_2 effects are simulated. Soybean responds positively to increased CO_2, but is the crop most affected by the high temperatures of the UKMO scenario.

The differences among countries in simulated crop yield responses to climate change without the direct effects of CO_2 are primarily related to differences in current growing conditions. Higher temperatures tend to shorten the growing period at all locations tested. At low latitudes, however, crops are currently grown at higher temperatures, produce lower yields, and are nearer the limits of temperature tolerances for heat and water stress. Warming at low latitudes thus results in accelerated growing periods for crops, more severe heat and water stress, and greater yield decreases than at higher latitudes. In many mid- and high-latitude areas where current temperature regimes are cooler, increased temperatures, while still shortening grain-filling periods, thus exerting a negative influence on yields, do not significantly increase stress levels. At some sites near the high-latitude boundaries of current agricultural production, increased temperatures can benefit crops otherwise limited by cold temperatures and short growing seasons, although the extent of soil suitable for expanded agricultural production in these regions was not studied explicitly. Potential for expansion of cultivated land is embedded in the BLS world food trade model and is reflected in shifts in production calculated by that model.

The GISS and GFDL climate change scenarios produced yield changes ranging from +30 to -30%. Effects under the GISS scenario are, in general, more adverse than under the GFDL scenario to crop yields in parts of Asia and South America, while effects under the GFDL scenario result in more negative yields in the United States and Africa and less positive results in former USSR. The UKMO climate change scenario, which has the greatest warming (5.2°C global surface air temperature increase), causes average national crop yields to decline almost everywhere.

Crop Yields with Adaptation

The adaptation studies conducted by the scientists participating in the project suggest that ease of adaptation to climate change is likely to vary with crop, site and adaptation technique. For example, at present, many Mexican producers can only afford to use small doses of nitrogen fertilizer at planting; if more fertilizer becomes available to more farmers some of the yield reductions under the climate change scenarios might be offset. However, given the current economic and environmental constraints in countries such as Mexico, a future with unlimited water and nutrients is unlikely. In contrast, switching from spring to winter wheat at the modelling sites in the former USSR produces a favourable response, suggesting that agricultural productivity may be enhanced there with the relatively easy shift to winter wheat varieties.

Adaptation Level 1, simulating minor changes to existing agricultural systems, compensated for the climate change scenarios incompletely, particularly in the developing countries. Adaptation Level 2, implying major changes to current agricultural systems, compensated almost fully for the negative climate change impacts in the GISS and the GFDL scenarios. With the high level of global warming as projected by the UKMO climate change scenario, neither Level 1 nor Level 2 adaptation fully overcame the negative climate change effects on crop yields in most countries, even when the direct CO_2 effects were taken into account.

Modelling the World Food System

The world food system consists of many actors, some powerful and others dependent. Producers and consumers interact through national and international markets. While there is a trend towards internationalization in the world food system, only ~15% of world cereal production currently crosses national borders. National governments shape the system by imposing regulations and by investments in agricultural research, improvements in infrastructure, and education. Although the system does not guarantee stability nor equity, food and fibre have been produced over time in increasingly efficient ways. These efficiencies have generated long-term real declines in prices of major food staples.

The Basic Linked System of National Agricultural Models

The Basic Linked System of National Agricultural Policy Models (BLS) consists of some 35 national and/or regional models: 18 national models, 2 models for regions with close economic cooperation (EU and Eastern Europe + former USSR [1]), 14 aggregate models of country groupings, and a small component that accounts for statistical discrepancies and imbalances during the historical period. The individual models are linked together by means of a world market module.

The political changes as well as changes in national boundaries of the very recent past are not captured in the BLS, although the model formulation has been adjusted, away from centrally planned economies to more market-oriented behaviour.

The BLS is a general equilibrium model system. This necessitates that all economic activities are represented in the model. Financial flows as well as commodity flows within a country and at the international level are consistent in the sense that they balance.

The country models are linked through trade, world market prices and financial flows. The system is solved in annual increments, simultaneously for all countries. It is assumed that supply does not adjust instantaneously to new economic conditions. Only supply that will be marketed in the following year is affected by possible changes in the economic environment. A first round of exports from all the countries is calculated for an initial set of world prices, and international market clearance is checked for each commodity. World prices are then revised, using an optimizing algorithm, and again transmitted to the national models. Next, these generate new domestic equilibria and adjust net exports. This process is repeated until the world markets are cleared in all commodities. At each stage of the iteration the domestic markets are in equilibrium. Since these steps are taken on a year-by-year basis, a recursive dynamic simulation results.

An upper bound on land available for cropping and for use as pasture is determined by the availability of land resources as well as economic conditions; e.g., by the economic returns to land. The physical resource limits in developing countries were derived from a FAO assessment of potential arable land. The responsiveness of how much land can be cultivated due to changing economic conditions is rather low since time and investment are needed to bring new land into cultivation.

Technological development is assumed to be largely determined by exogenous factors. Technical progress is included in the models as biological technical progress in the yield functions of both crops and livestock. Rates of technical progress were estimated from historical data and, in general, show a decline over time. Mechanical technical progress is part of the function determining the level of harvested crop area and livestock husbandry. Induced (endogenous) technical progress is not considered for any of these cases or for non-agricultural production.

Information generated by simulating with the BLS consists of a number of variables. At the world-market level these include prices, net exports, global production and consumption. At the country level, the information generated includes: producer and retail prices, level of production, use of primary production factors (land, labour and capital), intermediate input use (feed, fertilizer and other chemicals), level of human consumption, stocks and net trade, gross domestic product and investment by sector, population number and labour force, welfare measures such as equivalent income, and the level of policy measures as determined by the government.

BLS Reference Projections Without Climate Change

A set of reference scenarios was designed for studying the relative effects of climate change in relation to technology, population and economic growth. The standard reference scenario we describe, termed scenario REF-M, assumes 'business as usual' in the sense that no radical shifts in technological and political trends are included. Protectionist policy measures in agriculture, however, are assumed to be lowered to half the observed historical levels. The transition to such reduced protection of the agriculture sectors is implemented between the beginning of the simulation period and year 2020. Thereafter the respective policy settings are kept constant.

Another reference projection, scenario REF-H, assumes faster economic growth than the standard reference scenario. In the BLS, the dynamics of economic growth can be influenced by adjusting the rate of investment to the amount of technical progress. In scenario REF-H, growth of Gross Domestic Product (GDP) is especially higher for Asian countries compared to scenario REF-M. Similarly, a lower growth simulation run, scenario REF-L, has been created by curtailing investments in favour of consumption.

Additional reference projections with the BLS were based on the standard reference scenario REF-M. Two projections deal with the sensitivity of the world food system with respect to land development and availability of agricultural inputs. In scenario REF-MA, expansion of arable land after year 2000 is constrained to half the expansion in scenario REF-M. Another simulation run, scenario REF-MF, limited the use of fertilizers by implementing a tax on fertilizers of 50% in developed countries and of 33% in developing countries. In addition, a ceiling on fertilizer use per hectare was enforced. Both these scenarios were simulated to test the impact of possible agricultural policies geared towards reducing greenhouse gas emissions from agriculture (specifically, CO_2 release from deforestation and N_2O emissions from nitrogen fertilizers).

Finally, reference projection REF-MP uses the economic assumptions of the standard reference run REF-M but alters the demographic projections from medium population growth to low population growth.

All the simulations are carried out on a yearly basis from 1980 to 2060.

Assumptions About Population

Population growth rates were obtained from the UN World Population Prospects, Medium Variant in all simulations except for scenario REF-MP where the Low Variant was used. Since the UN projected national population levels only up to the year 2025, the remainder of the projection period was covered by growth rates compiled from long-term population projections of the World Bank. Labour participation rates are taken from projections of the International Labour Organization. The allocation of total labour force between agriculture and non-agricultural sectors responds to relative prices and incomes.

During the last two decades, population at the global level has been growing at an average rate of about 1.8% per annum, from 3 600 million in 1970 to about 5 300 million people in 1990. In the BLS standard reference projection, the average annual population growth rate is projected to decline gradually from 1.7% per annum during 1980 to 2000 to around 0.5% per annum for the period 2040 to 2060. This would bring global population numbers to 6 100 million in year 2000 and about 10 300 million in year 2060. Most of the increase in population numbers occurs in developing countries, increasing their share in world population from around 73% in 1980 to 78% in year 2000 and some 86% in year 2060. In 2060, almost 9 000 million people are projected to live in developing countries, of which approximately 5 000 million are in Asia, and 2 200 million in Africa.

Economic Growth

Growth rates in most of the national models of the BLS are determined based on three elements: (a) capital accumulation through investment and depreciation, related to a savings function that depends on lagged GDP levels as well as balance of trade and financial aid flows, (b) dynamics of the labour force as a result of demographic changes, and (c) technical progress. All regions, developed and developing alike, show a declining rate of GDP growth over time, which is consistent with historical developments. The declining rates of population growth (and the related decline in the growth of the labour force) as well as a general slowdown in productivity increases contribute to this development.

In the standard reference scenario, GDP at the global level increases at an average 2.4% annually during 1980 to 2020. Economic growth declines to about 1.5% per annum during 2020 to 2040. Overall, global GDP increases 4.4 times during the simulation period compared to a 2.4-fold increase in population numbers. This results in an average annual increase in GDP per caput of 1.4 and 1.1% in developed and developing countries, respectively, over the 80-year period from 1980 to 2060. It should be noted, however, that increases in per caput indicators are higher at regional and national levels compared to global figures owing to an aggregation effect induced by the demographic development, giving increasingly higher weights to poorer developing countries.

With faster economic growth, especially in developing countries such as in scenario REF-H, world GDP grows almost 5.6 times between year 1980 and 2060, compared to only 3.9 times as in the lower growth projection REF-L. Note that even with relatively minor variations on the basic economic assumptions, the resulting output in year 2060 varies by some 40%. Similarly, food production (measured in terms of net food energy, i.e., production less feed, seed and waste) as well as overall agriculture growth is projected to exceed population growth throughout the simulation period.

Production, Demand and Trade

The standard reference projection, scenario REF-M, presents the perspective of a world in which the effective demand for food grows substantially owing to higher incomes and larger populations. Technological progress and economic development assumed in the reference scenario allow this increase in demand to be met at somewhat decreasing world market prices for agricultural products, consistent with historical trends.

Global trade in the reference scenario increases somewhat faster than global agricultural production. For cereals, the share of net exports in global production is estimated to increase from 13% in 1980 to 15% in 2060, with wheat and coarse grains showing an almost three-fold and rice a four-fold increase in trade levels. In general, the share of global trade in global production of commodity aggregates increases gradually over time indicating a growing specialization in production. Increasing demand in developing countries, due to rising incomes and growing populations, leads to a deterioration in the level of agricultural self-sufficiency for this group of countries, which changes from a net surplus of about 3% in 1979/81 into a 1% deficit by the year 2060 caused by increasing deficits in cereals, meat and milk.

Risk of Hunger

Finally, we ask where all this leaves the hungry. To evaluate the impact of alternative scenarios on the poor in different countries, it was necessary to generate a consistent hunger indicator in the BLS. Country-wise estimates of the number of undernourished persons have been made by FAO. To recover the FAO method in a reduced form, suitable for use in the simulation models, a cross-country regression has been estimated explaining the share of people at risk of hunger by a measure of food energy availability relative to nutritional requirements. Food availability, in turn, depends on income and price levels.

In the standard reference scenario the incidence of hunger decreases markedly from an estimated 23% of population in developing countries (excluding China) in year 1980 to some 9% in year 2060. Yet, despite this remarkable improvement the estimated number of people at risk of hunger increases somewhat, from about 500 million [2] in 1980 to almost 720 million in year 2020, and some 640 million in the year 2060.

While the estimates show an improvement of the food security situation both in relative and absolute terms in Asian countries, the African continent experiences a mixed outcome largely due to the dramatic population increase. The estimated share of people at risk of hunger in total African population declines from 28% in year 1980 to 18% by 2060. However, the number of hungry is projected to increase more than three-fold, from about 120 million people in 1980 to 415 million in 2060, thereby making Africa the region with the largest number of undernourished. With lower population growth, scenario REF-MP, the estimated level of people at risk of hunger is reduced by more than 25%; higher economic development, scenario REF-H, reduces the number of hungry by some 40 million, i.e., approximately 10%.

Assessment of World food System under Alternative Scenarios

The evaluation of the potential impact of climate change on production and trade of agricultural commodities, in particular on food staples, is carried out by comparing the results of the climate change scenarios to the reference projections. Various aspects of these reference projections have been presented in the previous section.

The climate change yield impact scenarios devised within the project involve a large number of experiments that relate to:

1. different GCM doubled CO_2 simulations;
2. different assumptions with regard to impacts of climate change on plant growth and yield levels, such as physiological effects of 555 ppm CO_2, or time pace of impact;

3. different assumptions regarding farm-level adaptation to mitigate yield impacts;
4. policy changes to affect both the reference run and climate change experiments, e.g., population growth, trade policies, economic growth, and policies to reduce emission of greenhouse gases from agriculture, such as limitation of arable land expansion, rice cultivation, or use of chemical fertilizers.

Well over 70 experiments have been simulated. Results from six sets of simulation experiments are reported here:

1. Simulations *without* physiological effects of 555 ppm CO_2 on crop growth and yield.
2. Simulations *with* physiological effects of 555 ppm CO_2 on crop growth and yield.
3. Simulations with physiological effects of 555 ppm CO_2 on crop growth and yield, and adaptations to mitigate negative yield impacts at the farm-level that would not involve any major changes in agricultural practices.
4. Simulations with the physiological effects of 555 ppm CO_2 on crop growth and yield, and adaptations at the farm-level that, in addition to the former, would also involve major changes in agricultural practices.
5. Simulations with the physiological effects of 555 ppm CO_2 on crop growth and yield, but with temperature and precipitation changes projected by the GISS GCM for the 2030s.
6. Simulations with the GISS transient run A and associated gradual increases in CO_2 level.

Methods

Data on crop yield changes estimated for the different scenarios of climate change were compiled for 34 countries or major regions of the world. Most models included in the BLS distinguish between yield and acreage functions. The yield response functions of major crops use the level of fertilizer application and a term related to technology as explanatory variables.

While technical progress is specified relative to a time trend, the level of fertilizer application is derived from optimality conditions, i.e., by equating the marginal value product of fertilizer to its price. Yield variations caused by climate change were introduced into the yield response functions by means of a multiplicative factor applied to the relevant parameters in the mathematical representation. This implies that both average and marginal fertilizer productivity are affected by the imposed yield changes.

Alternative schemes for introducing yield changes are conceivable, e.g., with an additive term in the response functions rather than a multiplier. More empirical knowledge with regard to the effect of climate-induced yield changes on marginal productivity is needed to select the most appropriate implementation.

Since additional country and/or crop specific information to suggest explicit modifications of extents suitable for crop cultivation due to impacts of climate change was not available, the land allocation is only indirectly influenced through the implied changes in overall performance of the agricultural sector as well as changing comparative advantage of the competing crop production activities. It should be noted, however, that the BLS is equipped to handle explicit area constraints in the resource allocation module of the agricultural production component.

The adjustment processes taking place in the different scenarios are the outcome of the imposed yield changes triggering changes in national production levels and costs, leading to changes of agricultural prices in the international national markets. This in turn affects investment allocation and labour migration between sectors as well as reallocation of resources within agriculture.

Time is an important aspect in this assessment: the yield modifications due to climate change are assumed to start occurring in 1990, reaching their full impact in year 2060. This allows the economic actors in the national and international food system to adjust their behaviour over a 70-year period. Yet, the dynamic impacts in some of the scenarios are sizeable.

For the GISS transient scenario A, climate change yield impacts were phased in linearly between the climate 'snapshots', i.e., the yield change multiplier terms incorporated in the yield response functions of the BLS are being built up gradually as a function of time between 1990-2010, 2010-2030 and 2030-2050, so as to fully reach the specified impact levels respectively in years 2010, 2030 and 2050. Beyond year 2050 the yield change multiplier is extrapolated extending the trend of the period 2030-2050.

Climate Change Yield Impacts

Without Economic AdjustmentsBefore assessing the impacts of introducing a set of climate-change-induced yield modifications through simulation with the BLS, we may ask what distortion such a change in agricultural productivity would imply for the world food system. This measure of distortion is termed the *static* climate change yield impact, as it measures the hypothetical effect of yield changes without adjustments of the economic system taking place over time. It refers to a state of the system that is *not* in equilibrium. As such it is only of theoretical interest, but helps in the understanding and quantification of the nature and magnitude of adjustments taking place due to changing economic conditions.

The estimates of*static* climate change yield impacts, without assuming direct physiological effects of 555 ppm CO_2 on crop growth and yields, represent a fairly pessimistic outlook, with decreases in crop productivity on the order of 20 to 30%. Such an assumption is not regarded as very probable and will not be further discussed in the analysis.

When direct physiological effects of CO_2 on yields are included, the magnitude and even the direction of the aggregate *static* impact at the world level varies with GCM climate scenario and with the assumptions regarding farm-level adaptation. In all cases the most negative effects are obtained in scenarios using the UKMO climate change scenario, which has the highest mean global warming, 5.2°C. Results derived from the GISS scenario show only small negative effects or even gains at the global level.

Climate Change Yield Impacts with Economic Adjustments

The calculations above discuss an effect that would result if climate-induced yield changes were to occur without agronomic and economic adjustment. In the scenario assumptions, however, yield productivity changes are introduced gradually to reach their full impact only after a 70-year period, from 1990 to 2060. In scenarios with shortfalls in food production caused by climate change, market imbalances push international prices upwards and provide incentives for reallocation of capital and human resources. At the same time, consumers react to price changes and adjust their patterns of consumption.

When the physiological effects of 555 ppm CO_2 on yields are included in the assessment, cereal prices increase on the order of 24 to 145% relative to the standard reference projection. The index of crop prices increases by 8 to 90%, depending on GCM climate change scenario. Changes caused by GISS and GFDL scenarios are rather modest, resulting in an increase of about 25 to 33% in cereal prices, and an increase of less than 20% in overall crop prices. Only the yield impacts derived under a climate change as projected by the UKMO GCM produced large agricultural price increases. On the other hand, under the GISS-A climate scenario, where impacts are dominated by positive physiological effects of CO_2, major price decreases occur.

Price changes are further reduced when farm-level adaptation is considered. The crop price index rises less than 10% in both GISS and GFDL simulation runs. The UKMO projection, however, still produces a two-thirds crop price increase. A. Note that in the GISS doubled CO_2 and GISS transient run, positive crop impacts dominate for about half the simulation period. In the GISS transient run the crop price index falls initially by as much as 10% below the level of the reference run. Then, as negative impacts in developing countries increase and beneficial impacts in developed countries level off, prices return to and finally exceed the price index of the reference scenario.

With adaptation measures involving major changes in agricultural practices, i.e., Adaptation Level 2, prices would even fall below reference run levels in the GISS and GFDL scenarios. Note that the assumptions underlying Adaptation Level 2, sometimes requiring major investments, may not be economically viable. Scenarios with low-cost adaptation measures, i.e., Adaptation Level 1, appear to be more realistic.

It is important to note that these changes in comparative advantage between developed and developing regions are likely to accentuate the magnitude of the static impacts suggested by the analysis without economic adjustment. Winners are likely to gain more, and losers to lose even more. We can distinguish two prototypical situations in these scenario results. (1) When global supply is only marginally affected, there is little impact on prices. Then the shift in relative productivity from developing to developed regions dominates the adjustment process. For instance, in the GISS and GFDL scenarios with farm-level adaptation, agriculture production shifts somewhat from developing to developed countries taking account of the differences in projected yield changes. (2) When global crop yields are strongly affected, as in the UKMO scenario, the supply gap is so substantial that massive price increases result. These in turn provide production incentives to both regions to recover more than half the production forgone due to climate change (according to static crop model estimates).

Global wheat production is less likely to suffer negative climate impacts than other cereal commodities; it increases in GISS and GFDL scenarios and declines less than other cereals in the UKMO scenario. This is likely due to the location of major wheat-producing regions at mid and high latitudes, where yield declines are projected to be lower.

Net imports of cereals into developing countries increase under all scenarios. The change in cereal imports, relative to the standard reference projection, is largely determined by the magnitude of the estimated yield change, the change in relative productivity in developing and developed regions, the change in world market prices, and changes in real incomes of consumers in developing countries. For example, under the GISS climate change scenarios, productivity is altered in favour of developed countries with relatively small changes in incomes and prices, resulting in pronounced increases of net cereal imports into developing countries.

With less agricultural production in developing countries and higher prices on international markets, the estimated number of people at risk of hunger is likely to increase. This occurs in all but one scenario. The largest increase is to be expected from the UKMO scenario without CO_2 physiological effects; the smallest change, a decline of 2%, occurred in the GISS scenario considering physiological effects of increased CO_2 and Adaptation Level 2.

The distortions of the world food system simulated in the climate change scenarios fall well within the range of estimates obtained from the different reference projections. For instance, the decline in cereal production even under the worst climate change scenario based on the UKMO GCM experiment (assuming physiological effects of increased atmospheric CO_2 concentrations and some farm-level adaptation), amounts to less than half the difference between the cereal production levels simulated in the higher and lower economic growth reference projections, scenarios REF-H and REF-L. However, the ability of the world food system to absorb negative yield

The effects of changes in climate on crop yields are likely to vary greatly from region to region across the globe. The results of the scenarios tested in this study indicate that the effects on crop yields in mid- and high-latitude regions appear to be positive or less adverse than those in low-latitude regions, provided the potentially beneficial direct physiological effects of CO_2 on crop growth can be fully realized. From a development perspective, the most serious concern relates to the apparent difference in incremental yield impacts between developed and developing countries. The scenario results suggest that if climatic change were to retard economic development beyond the direct effects on agriculture in the poorer regions, especially in Africa, then overall impacts could be sizeable.

In all climate change scenarios, relative productivity of agriculture changes in favour of developed countries, with implications on resource allocation. Economic feedback mechanisms are likely to emphasize and accentuate the uneven distribution of climate change impacts across the world, resulting in net gains for developed countries in all but the UKMO scenarios and a noticeable loss to developing countries. As a result, net imports of cereals into developing countries increase in all scenarios, on the order of 20 to 50% compared to trade in the reference scenario.

Including direct physiological effects of CO_2 on crop yields, world cereal production is estimated to decrease between 1 to 3% under GISS and GFDL scenarios, and 7% in the UKMO climate scenario based projections. Assuming adaptation to climate change at farm-level, cereal production would still be reduced between 0 to 2% and 5%, respectively, for the GISS/GFDL and UKMO scenarios. The largest negative changes would occur in developing countries, averaging around -10%. This loss of production in developing countries, together with rising agricultural prices, is likely to increase the number of people at risk of hunger, in the order of 5 to 15 % in the less severe climate scenarios, and ~50% in the UKMO based projections. Under a possibly more realistic climate change scenario with a lower climate sensitivity to increasing greenhouse gas concentrations, aggregate crop productivity at global level increases by ~10% until year 2060, i.e., the end of the simulation period considered in the analysis. Impacts are assessed to be positive in almost all of the 10 aggregate world regions reported in the study. However, the simulated percentage increases in developed countries are about twice the increases in developing regions.

The analysis also shows that the yield impacts do not only vary with geographic region, but are also unequally distributed over time. Results of scenarios based on the GISS transient scenario A demonstrate that benefits from physiological effects due to increasing atmospheric CO_2 levels may outweigh negative impacts from changing temperature and precipitation regimes at least in the near-term. The yield-increasing factors in that scenario dominate possible negative impacts until year 2020. Understanding the biophysical processes of CO_2 and climate change effects on crops remains an important research area.

It must be realized, however, that the ability to estimate climate change yield impacts on world food supply, demand and trade is surrounded by large uncertainties regarding important elements, such as the magnitude and spatial characteristics of climate change, the range and efficiency of adaptation possibilities, the long-term aspects of technological change and agricultural productivity, and even future demographic trends. Also, the adoption of efficient adaptation techniques is far from certain. In developing countries there may be social, economic or technical constraints, and adaptive measures may not necessarily result in sustainable production over long time-frames.

Determining how countries, particularly developing countries, can and will respond to reduced yields and increased costs of food is a critical research need arising from this study. Will such countries be able to import large amounts of food? Will the burden for adaptation be passed on to the poorest? From a political and social standpoint, the results of the study indicate the potential for a decrease in food security in developing countries. The study suggests that the worst situation arises from a scenario of severe climate change, low economic growth, continuing large population increases, and little farm-level adaptation. In order to minimize possible adverse consequences, like production losses, food price increases, environmental stresses, and an increase in the number of people at risk of hunger, the way forward is to encourage the agricultural sector to continue to develop crop breeding and management programmes for heat and drought conditions, in combination with measures taken to preserve the environment, to use resources more efficiently, and to slow the growth of the human population of the world. The latter step would also be consistent with efforts to slow emissions of greenhouse gases, and thus the rate and eventual magnitude of global climate change.

In the face of these uncertainties, both national and international organizations should encourage the development of new approaches likely to be effective in preparing for climate change. Agricultural research would benefit from increased attention to both macroclimate and microclimate in all experiments and variety trials. Another climate change impact potentially significant for future agricultural production is soil organic matter loss due to soil warming. Considering the vulnerability of agricultural production to the occurrence of climate extremes, research should be directed to determine what are the heat-tolerance limits of currently grown and of alternative crops and varieties. At what threshold values of air or soil temperature do severe problems begin? What agronomic methods are the best to moderate the thermal regime affecting crop growth? To the extent that the progressive greenhouse effect cannot be prevented in practice, policies should be devised to facilitate the adjustment of agriculture to the likelihood of environmental change. Such adjustments may include modification of agronomic practices, adoption of crops known to be heat-resistant and drought-resistant, increased efficiency of irrigation and water conservation, and improved pest management. Such adjustments are worthy of being implemented in any case, be it with or without climatic change.

Although some countries in the temperate zone may reap some benefits from climate change, many countries in the tropical and subtropical zones appear to be more vulnerable. Particular hazards are the possibly increased flooding of low-lying areas, the increased frequency and severity of droughts in semi-arid areas, and potential decreases in attainable crop yields. It happens that the latter countries tend to be the poorest and the least able to make the necessary economic adjustments. Much of the expected change in global climate is due to the past and present activities of the industrial countries; so it is their responsibility to commit themselves to, and to play an active role in, a comprehensive international effort to prepare for the likely consequences.

References

Parry, M.L., Carter, T.R. and Konijn, N.T. (eds.). 1988. *The Impact of Climate Variations on Agriculture. Vol. 1: Assessments in Cool Temperate and Cold Regions. Vol. 2: Assessments in Semi-arid Regions.* Kluwer, Dordrecht.

Pearman, G. 1988. *Greenhouse: Planning for Climate Change.* CSIRO, Canberra.

oriented wheat yield model. In: *ARS Wheat Yield Project.* W.O. Willis (ed.). Department of Agriculture, Agricultural Research Service. ARS-38. Washington DC.

Rosenzweig, C. and Iglesias, A. 1994. *Implications of Climate Change for International Agriculture: Crop Modeling Study.* US Environmental Protection Agency. EPA 230-B-94-003, Washington DC.

Smith, J.B. and Tirpak, D.A. (eds.). 1989. *The Potential Effects of Global Climate Change on the United States.* Report to Congress. EPA, Washington DC.

9

Climate Change and Global Agricultural Vulnerability

The potential impacts of climate change on agriculture are highly uncertain. The large number of studies conducted over the past few years for many different sites across the world show few, if any, robust conclusions of either the magnitude or direction of impact for individual countries or regions. Where apparent consensus exists it frequently appears to occur because only one or two studies have been conducted using a single climate scenario. Many such studies have focused on doubled (2xCO_2,) General Circulation Model (GCM) equilibrium scenarios. These do not begin to describe the variety of climatic conditions any particular region is likely to experience as the actual climate changes over time.

Potential future climate changes are also made more uncertain because of the recently recognized role of sulphate aerosols which may partly offset the warming expected from increased concentrations of CO_2, methane, nitrous oxide and other radiatively active trace gases. The significant spatial variation in sulphate aerosol concentrations means that the regional pattern of climate change may be quite different from that simulated on the basis of CO_2 increase alone. The short lifetime of aerosols in the atmosphere (a few days) means that if the use of high sulphur coal in India or China increases or efforts to control sulphur emission in the United States or Europe are intensified, the spatial pattern of climate change could change significantly within a relatively short period of time due to changes in the aerosol cooling effect.

Different impact methodologies also yield widely varying results of the direct impacts of climate change on crop yields and agricultural production even when examining the same region and the same climate scenarios. The socio-economic environment, agricultural technology and natural resource base will also necessarily undergo profound changes over the next 100 years whether agriculture meets the many challenges of feeding the world's growing population or fails to do so.

The robust conclusion that does emerge from impact studies is that climate change has the potential to change significantly the productivity of agriculture at most locations. Some currently highly productive areas may become much less productive. Some currently marginal areas may benefit substantially while others may become unproductive. Crop yield studies show regional variations of +20, 30 or more per cent in some areas and equal size losses in other areas. Most areas can expect change and will need to adapt, but the direction of change, particularly of precipitation, and required adaptations cannot now be predicted. Nor may it ever be possible to predict them with confidence. Current evidence suggests that poleward regions where agriculture is limited by short growing seasons are more likely to gain while subtropical and tropical regions may be more likely to suffer drought and losses in productivity. However, these broad conclusions hardly provide the basis for mapping out a long-term strategy for agricultural adaptation. Thus, policy must retain flexibility to respond as conditions change.

A further issue is how do climate change impacts on agricultural production fit within the other pressing challenges facing agriculture in different regions of the world. Is climate change a minor threat, likely to be undetected among the many changes that will reshape the agricultural sectors of the world's economies? Or is it another critical challenge to an agricultural sector straining to cope with a growing population, resource degradation, tighter constraints on available resources, and exhaustion of technological capabilities to expand production using existing land and water resources?

It is useful to place some of the 2xCO_2 agricultural projections in the context of other future projections. If we accept long-term demographic trends, the largest absolute addition to the world's population will occur during the decade of the 1990s, the growth rate having already slowed from that of the 1950s and 1960s. By the time 2xCO_2 climate scenarios are expected to be realized (some time around 2100 or later), the world population will have stabilized and agricultural research will no longer face the challenge of increasing productivity to keep up with a growing population.

Therefore there is a need for more specific analysis about how climate will change over the next 10, 20 or 30 years rather than over the next 100. It also provides a caution not to consider our response to climate change apart from our response to the immediate needs of agriculture: feeding a growing population where presently an estimated 740 million people still suffer from hunger and malnutrition while maintaining the productivity of basic agricultural resources and meeting the demands placed on agriculture to minimize damage to the environment.

Impact Assessment Methodologies

Climate change presents a challenge for researchers attempting to quantify its impact due to the global scale of likely impacts, the diversity of agriculture systems, and the decades' long time scale. Current climatic, soil and socio-economic conditions vary widely across the world. Each crop and crop variety has specific climatic tolerances and optima. It is not possible to model world agriculture in a way that captures the details of plant response in every location. The availability of data with the necessary geographic detail is currently the major limitation rather than computational capability or basic understanding of crop responses to climate. A specific problem has been how to take the detailed knowledge of plant response into aggregate assessments of regional assessments. In general, compromises are necessary in developing quantitative analyses at regional scales.

There are two basic approaches for evaluating crop and farmer response to changing climate: (1) structural modelling of the agronomic response of plants and the economic/management decisions of farmers based on theoretical specifications and controlled experimental evidence; and (2) reliance on the observed response of crops and farmers to varying climate.

For the first approach, sufficient structure and detail are needed to represent specific crops and crop varieties for which responses to different conditions are known through detailed experiments. Similar detail on farm management allows direct modelling of the timing of field operations, crop choices, and how these decisions affect costs and revenues. These approaches typically model a representative crop or farm. Both in the case of economic models of farm decisions and in the case of crop response models, the original purpose was to improve understanding of how the crop grows or how a farmer manages. In the case of models of a representative farm, one might hope to offer prescriptive advice for the farmer - where farm operations differ from the profit maximizing (or cost minimizing) model results, it provides guidance for how farmers might improve farm performance. In both cases, the idealized representation of the crop and farm operation tends to give results that differ markedly from the actual experience on farms operating under real world conditions. This may reflect the fact that farmers do not operate as profit maximizers (they could improve their performance) or that the models fail to consider some of the factors that the farmer takes into account such as risk or lack of immediate employment alternatives. Because of the idealized nature of these models, many analysts consider that these provide evidence of the *potential production or potential profitability.* Imposing climate change on these models gives estimates of how potential production may change due to climate change. Using these results as indicative of how climate will actually affect agriculture thus rests on the assumption that the change in the potential represents the change likely to be actually experienced. Many approaches of this type have used detailed crop response models requiring daily weather records. For aggregate analyses inferences must be made from relatively few sites and crops to large areas and diverse production systems because of the complexity of the models and the need for detailed data on weather over a decade or more.

The work of Leemans and Solomon, is in a similar vein, choosing much simpler representations of crop/climate interactions but is still related to basic agronomic representation of crop growth in response to temperature and precipitation. The advantage of their approach is that, because of the minimal amounts of climatic data required (mean monthly data on temperature and precipitation), the crop models can be applied at a resolution of 0.5° x 0.5° latitude-longitude grids.

The second approach, relying on observed response of crops and farmers, provided some of the earliest estimates of the potential effects. The simplest example of this approach is to observe the current climatic boundaries of crops and to redraw these boundaries for a predicted changed climate. In a similar vein, researchers have applied statistical analysis of data across geographic areas to separate climate from other factors (e.g., different soil quality, varying economic conditions) that explain production differences across regions and have used these to estimate the potential agricultural impacts of climate change. An advantage of using direct evidence from observed production is that the data reflect how farmers operating under commercial conditions and crops growing under such conditions actually respond to geographically varying climatic conditions. Here, the most recent work uses extremely reduced form models although estimation of more detailed structural models is possible. Darwin *et al.*, use revealed evidence from geographic variation in climate in a global model, allocating production and input use to climatically determined land classes based on current production patterns. Climate change impacts are then simulated by altering the distribution of land classes and assuming that when an area's land class changes, its underlying production level changes to that of the new land class.

The Darwin *et al.*, approach links the basic agricultural productivity of land classes, described by a production function, with a computable general equilibrium model of the world economy. Thus, actual production in a region or land class depends on the final market clearing prices. The model also treats interactions with other sectors of the economy, most importantly sectors that compete for land and water.

Crops Response Estimates for Different Regions of the World

The wide ranges of estimates are due to several, as yet unresolved, factors. Differences among climate scenarios are important and these can generate wide ranges of impacts even when using identical methods for the same regions. For example, a study of the potential impact on rice yields conducted for most of the countries of South and Southeast Asia and for China, Japan, and Korea using the same crop model found yield changes for India to range from -3 to +28%, for Malaysia from +2 to +27%, for the Philippines from -14 to +14%, and for mainland China from -18 to -4% depending on whether the GISS, GFDL or UKMO climate scenario was used.

The impacts across sites can vary widely within a region. Thus, how many and which sites are chosen to represent a region and how the site-specific estimates are aggregated can have important effects on the results. Studies for the United States and Canada demonstrate the wide range of impacts across sites with total or near-total crop failure every year projected for wheat and soybeans at one site in the United States but wheat yield increases of 180 to 230% for other sites in the United States and Canada.

Whether and how changes in a crop variety are specified in a study can have a large impact. Studies conducted of wheat response in Australia found impacts ranging from -34 to +65% for the same climate scenario and site depending on which known and currently grown wheat cultivar was specified in the crop model. Similarly, Matthews *et al.* concluded that the severe yield losses in South, Southeast and East Asia for rice in many scenarios was due to a threshold temperature effect that caused spikelet sterility but that genetic variation with regard to the threshold likely provided significant opportunity to switch varieties as temperatures rose. Thus, an impact analysis that narrowly specifies a crop variety is likely to generate a much different estimated impact than an analysis that specifies responses on the basis of the genetic variation across existing cultivars. Some studies have attempted to evaluate how future crop breeding may change the range of genetic variability available in future varieties.

Finally, the estimated amount of adaptation likely to be undertaken by farmers varies. Fundamental views about how the farm sector responds to changing conditions (of any kind) shape the choice of methodological approach, and these methodological approaches can give apparently widely different estimates of impact. Specification of the crop variety in a crop response model illustrates this difference. For some analysts, the prospect that farmers will not change the variety of crop grown over the next 100 years as climate, technology, prices and other factors change is so remote that they choose to represent change among varieties as an essentially autonomous response of the farm sector. Other analysts choose more specific crop variety characteristics, viewing even crop variety change as neither automatic nor without cost. For example, different varieties of wheat produce flours with different characteristics and the cultural practices for growing spring and winter wheat differ. Similarly, studies of impacts on Japanese rice production estimate negative impacts for the southern parts of the country because of the climate tolerances of Japonica rice which is preferred over Indica varieties in Japan.

The differences from simply whether or not one assumes farmers will adopt the better adapted variety are large but these differences are potentially magnified many fold because the series of potential adaptations are broad with some requiring more specific recognition, action and investment by farmers. How do farmers choose a planting date - by planting at the same time each year regardless of weather conditions or by planting when soil temperatures are sufficient for crop growth, when the rainy season starts, or when the fields can be tilled? If the decision is partly keyed to weather conditions then the farm decision-making process will lead to some amount of autonomous adjustment to climate change. Similarly, will the changes in tillage and irrigation practices, crop rotation schemes, crops, and crop processing and harvesting that are likely to occur over the next 100 years due to many factors also reflect changes in climate that are occurring simultaneously, or will farmers be unable to detect climate change and therefore fail to adapt these systems, becoming and remaining ill-adapted to the climate conditions occurring locally? If they adapt to current conditions (but cannot confidently look ahead) how maladjusted will their long-lived investments be after 3, 5, 10 or 20 years of continuous changes in climate?

Global Studies and their Implications for Regional Effects

Accurate consideration of national and local food supply and economic effects depends on an appraisal of changes in global food supply and prices. International markets can moderate or reinforce local and national changes. In 1988, for example, drought presented a more severe threat because it occurred coincidentally in several of the major grain-growing regions of the world. Reilly *et al.,* demonstrate that considering country-level production impacts of climate change in the absence of consideration of the global impacts can generate highly misleading results. Agricultural exporting countries, whose productivity is reduced by climate change, may find themselves with a financial bonanza if world agricultural prices rise because of climate change. These same countries may suffer significant economic loss if climate change turns out to be generally beneficial to world agriculture even if agricultural productivity in their country benefits. This feature of the agricultural economy is well-known and reflects what is, in aggregate, an inelastic demand for food. This point, which is a fundamental observation of agricultural economists, means that absolutely no implications for food availability, price or farm financial success can be drawn from local and country-level estimates of production impacts of climate change unless one assumes that production changes around the world will generally balance to leave little impact on global production and prices. A country may attempt to carry out a set of policies that maintains a neutral effect on the country's agricultural sector *vis-a-vis* the rest of the world, but maintaining such policies will generally entail significant economic cost through subsidization of domestic agricultural production and/or consumption or through import or export controls. There are many different ways these costs may be borne (higher food prices, government expenditures, lost efficiency in the producing sector, lost export opportunities) depending on how the policies are structured.

There are now a number of different attempts to estimate the impacts of climate change on global agriculture, in part to consider the global impacts but more importantly to consider more accurately what the regional impacts could be recognizing that what happens to global agriculture due to climate change will likely be more important for the viability and economic success of local agriculture than what happens to local production potential itself. Kane *et al.*, and Tobey *et al.*, examined the sensitivity of agriculture to potential yield losses in major temperate grain-growing regions based on very stylized climate change impacts. They loosely linked the potential for yield losses in temperate regions to climate projections that showed increasing aridity in the continental mid-latitude areas. They made alternative assumptions about how agriculture might be affected in higher latitudes and in the tropics. They also developed scenarios that reflected the estimated yield impacts for different parts of the world that were summarized in the 1990 Intergovernmental Panel on Climate Change assessment. The yield response estimates used by Rosenzweig and Parry, also reported in Fischer *et al.* were also the basis of Reilly *et al.*, and in greater detail Reilly *et al.*,. Many of the general conclusions are similar between the studies indicating that, given a set of yield shocks, economic modelling of international markets in itself is not a major source of difference in results even though there are major differences in the modelling approaches. Rather these different economic modelling approaches focus on different aspects and degrees of detail of agricultural economic interactions among crops, livestock, land use and the rest of the economy.

Among the issues that give rise to uncertainty in these studies are the following factors:

1. The timing of expected climate change. For example, Rosenzweig and Parry, assume that the 4.0 to 5.2°C scenarios occur in 2060, but the most recent IPCC work suggests the mean estimate for 2060 is closer to 1.5°C and that the range of global temperature impacts by 2100 is likely to be between 1.0 and 5°C.
2. Aggregation from detailed sites. Detailed plant growth models, the basis for many studies, require daily temperature and precipitation records

for a 10- to 30-year historical climate record and detailed soil data, limiting the number of sites for which data are readily available and that can be practicably assessed. An alternative approach makes use of geographic information system databases that contain more extensive information on current climates across the world. These efforts have not been linked to an economic model. Results confirm the pattern of relative decreased crop potential in the tropical areas and increased potential in the northern areas but are not aggregated to determine the net global effect.

3. Coverage of agricultural activities. Simulation of crop response models has been limited to a few major crops for a region, usually important grain crops, with yield effects extended to other crops. Left out are the indirect impacts of climate change through impacts on insect, disease and weed pests; on soils; and on livestock production. Mendelsohn *et al.*, argue that their statistical approach accounts for all agricultural activities, implicitly accounting for the full effects of climate.
4. Other resource changes and competition for resources from other sectors. Land and water resource allocation is a conspicuous limitation in global studies. Water demand for other uses will grow, water use may have reached or passed sustainable levels of use in some areas, irrigation is responsible for salinization and land degradation, and water pricing and water system management are far from efficient under current conditions. Climate change also will affect demand for resources from other sectors.

The Darwin *et al.*, study addresses many of these considerations in a global model including eight world regions. An Applied Computable General Equilibrium Model (ACGE), land and climatic resource changes are based on a geographic information system; changing climate shifts the distribution of land across several agroclimatic land classes. Other resource-using sectors are included and are also affected by climate change. The model is a static one, imposing climate change on current economic and agricultural markets and thus does not directly address the issue of timing.

The global results are comparable to Rosenzweig and Parry in terms of direct supply impacts for the world in the 'no adaptation' case, but the study finds that adaptation is able to turn global losses to small global benefits (unrestricted case). Even when the model is constrained to continue to produce on existing amounts of land within each region and prices are not allowed to respond, adaptation mitigates a significant share of the losses. These results contrast with those of Rosenzweig and Parry,, giving generally smaller impacts and possible benefits even without the CO_2 effect and in that they show adaptation to be quite important.

Again, the global results are important because they are the first step in considering whether a local economy's consumers will be able to purchase food if it is unavailable domestically, how local producers may be affected by changes in demand for their crops, or how the cost of a country's agricultural policies may change because of changing international market conditions.

Regional vulnerability

The previous sections documented the wide range of uncertainty in the potential direction and magnitude of climate change impact. While many new studies have been conducted, most have focused on specific climate scenarios associated with $2xCO_2$ GCM scenarios or arbitrary changes in climatic conditions to provide evidence of the general sensitivity of agriculture and crop production to climate change. The wide range of estimates limits the ability to extend, interpolate or extrapolate from the specific climate scenarios used in these studies to 'more' or 'less' climate change or to draw implications for impacts beyond the sites where studies were conducted.

Given these uncertainties in both magnitude and direction of impact, a key issue is *vulnerability* to possible climate change. Vulnerability is used here to mean the *potential* for negative consequences that are difficult to ameliorate through adaptive measures given the range of possible climate changes that might reasonably occur. Defining an area or population as vulnerable is, thus, not a prediction of negative consequences of climate change; it is an indication that across the range of possible climate changes, there are some climatic outcomes that would lead to relatively more serious consequences for the region than for other regions.

Farmer or farm sector vulnerability may be measured in terms of impact on profitability or viability of the farming system. Farmers with limited financial resources and farming systems with few adaptive technological opportunities available to limit or reverse adverse climate change may suffer significant disruption and financial loss for relatively small changes in crop yields and productivity or these farms may be located in areas more likely to suffer yield losses. For example, Parry *et al.* focused on semi-arid and cool temperate and cold agricultural areas as those that might be more clearly affected by climate change and climate variability.

Regional economic vulnerability reflects the sensitivity of the regional or national economy to farm sector and climate change impacts. A regional economy that offers only limited employment alternatives for workers dislocated by the changing profitability of farming and other climatically sensitive sectors may be relatively more vulnerable than those that are economically diverse. For example, Rosenberg, examined the Great Plains area of the United States because of its heavy dependence on agriculture. Increasing aridity is expected in this region under climate change and thus it was considered to be potentially more economically vulnerable than other regions in the USA.

Hunger vulnerability has been used to mean the 'aggregate measure of the factors that influence exposure to hunger and predisposition to its consequences' involving 'interactions of climate change, resource constraints, population growth, and economic development'. Downing, concluded that the semi-extensive farming zone, on the margin of more intensive land uses, appears to be particularly sensitive to small changes in climate. Socio-economic groups in such areas, already vulnerable in terms of self-sufficiency and food security, could be further marginalized. We likely ought not to look only at agriculturally dependent people. The means people have within society and the family to obtain food and how their allocation will change if production potential changes must be considered. A poor urban household may suffer due to production losses elsewhere in the region while the rural farmer may continue to eat. Or, women and children of rural peasant farms may go hungry, while 'excess' production from the region is sold. Assessing who has the means and rights to food during shortfalls is thus likely more critical in a climate vulnerability study than assessing how production may change.

Given the diverse currently existing conditions, the geographical variation likely to exist in any climate change scenario, and the wide uncertainty that must be associated with local prediction of future climates, some vulnerable agricultural areas and populations likely exist for nearly every region even if the expected value for the region is a net benefit. This makes vulnerability a relative concept - while there may be a few areas where even the most extreme climate change we can imagine would not generate losses, in general, the problem is to consider whether a particular region or population is relatively more vulnerable than others.

While perhaps most difficult to evaluate, vulnerability in terms of hunger and malnutrition ought to be the first concern. If so, then we can almost certainly eliminate the richer countries of the world as vulnerable. For poorer regions, it is the poorest members of these areas or those that could be made poor by climate change that are most at risk. The wide uncertainty with regard to local and regional climate change means it is difficult to rule out negative possibilities for any area. Thus, without even considering specific climate scenarios, we can assert that those who are currently poor, malnourished and dependent on local production for food are the most vulnerable in terms of hunger and malnutrition to climate change of the world's populations. Similarly, severe economic vulnerability is most likely where a large share of the population depend on agriculture, leaving little alternative employment opportunities. Again, we need not assess climate scenarios or projected yield changes to establish where these vulnerable populations live.

Because of the wide range of uncertainty in precipitation, the only climatic dimension likely to enter significantly in an assessment of vulnerability is temperature. Cool regions are more likely to be limited by low temperatures and thus warming may prove beneficial - these areas may still suffer if precipitation changes are adverse. But, further warming is unlikely to benefit already warm regions. Thus, global warming appears somewhat stacked against the already warm areas. Coincidentally (or not), these regions tend to also be home to some of the world's poorest.

The focus on hunger and malnutrition as a first priority does not mean that other types of vulnerability are unimportant. Regional economic development, land degradation, or increased environmental stress resulting from agricultural production under a changed climate are important concerns as well.

Adaptation Potential and Policies

The hierarchy of damage considerations as above - hunger, regional economic, farmer/farm sector, and yield vulnerability - helps to focus on adaptive strategies that reduce vulnerability. How can we avoid yield failures? If yields fail, what other crops can be grown? If farming becomes uneconomic, what other opportunities for employment exist? If the people of the region can no longer produce food, what other sources of food are available and how will they earn the income necessary to purchase food or what other means does the society in which they live have to provide food assistance?

Historically, farming systems have adapted to changing economic conditions, technology and resource availabilities and have kept pace with a growing population. Evidence exists that agricultural innovation responds to economic incentives such as factor prices and can relocate geographically. A number of studies indicate that adaptation and adjustment will be important to limit losses or to take advantage of improving climatic conditions.

Despite the successful historical record, issues of future adaptation to climate change arise with regard to whether the rate of change of climate and required adaptation would add significantly to the disruption likely due to future changes in economic conditions, technology and resource availabilities. If climate change is gradual, it may be a small factor that goes unnoticed by most farmers as they adjust to other more profound changes in agriculture stemming from new technology, increasing demand for food, and other environmental concerns such as pesticide use, water quality, and land preservation. However, some researchers see climate change as a significant addition to future stresses; where adapting to yet another stress such as climate change may be beyond the capability of the system. Part of the divergence in views may be due to different interpretations of adaptation which include: the prevention of loss, tolerating loss, or relocating to avoid loss. And, while the technological potential to adapt may exist, the socio-economic capability to adapt likely differs for different types of agricultural systems.

The Technological Potential to Adapt

Nearly all agricultural impact studies conducted over the past five years have considered some technological options for adapting to climate change. Among those that offer promise are:

— *Seasonal changes and sowing dates.* For frost-limited growing areas (i.e., temperate and cold areas), warming could extend the season, allowing planting of longer maturity annual varieties that achieve higher yields. For short-season crops such as wheat, rice, barley, oats, and many vegetable crops, extension of the growing season may allow more crops per year, autumn planting, or, where warming leads to regular summer highs above critical thresholds, a split season with a short summer fallow may be possible. For subtropical and tropical areas where growing season is limited by precipitation or where the cropping already occurs throughout the year, the ability to extend the growing season may be more limited and depends on how precipitation patterns change. A study for Thailand found yield losses in the warmer season partially offset by gains in the cooler season.

— *Different crop variety or species.* For most major crops, varieties exist with a wide range of maturities and climatic tolerances. For example, Matthews *et al.* identified wide genetic variability among rice varieties as a reasonably easy response to spikelet sterility in rice that occurred in simulations for South and Southeast Asia. Studies in Australia showed that responses to climate change are strongly cultivar dependent. Longer-season cultivars were shown to provide a steadier yield under more variable conditions. In general, such changes may lead to higher yields or may only partly offset losses in yields or profitability. Crop diversification in Canada and in China has been identified as an adaptive response.

— *New crop varieties.* The genetic base is broad for most crops but limited for some (e.g., kiwi fruit). A study by Easterling *et al.*, explored how hypothetical new varieties would respond to climate change. Heat, drought and pest resistance; salt tolerance, and general improvements in crop yield and quality would be beneficial. Genetic engineering and gene mapping offer the potential for introducing a wider range of traits. Difficulty in assuring traits are efficaciously expressed in the full plant, consumer concerns, profitability and regulatory hurdles have slowed the introduction of genetically engineered varieties compared with early estimates.

— *Water supply and irrigation systems.* Across studies, irrigated agriculture is in general less negatively affected than dryland agriculture but adding irrigation is costly and subject to the availability of water supplies. Climate change will also affect future water supplies. There is wide scope for enhancing irrigation efficiency through adoption of drip irrigation systems and other water-conserving technologies but successful adoption will require substantial changes in how irrigation systems are managed and how water resources are priced. Because inadequate water systems are responsible for current problems of land degradation, and because competition for water is likely to increase, there likely will be a need for changes in the management and pricing of water regardless of whether and how climate changes. Tillage method and incorporation of crop residues are other means of increasing the useful water supply for cropping.

— *Other inputs and management adjustments.* Added nitrogen and other fertilizers would likely be necessary to take full advantage of the CO_2 effect. Where high levels of nitrogen are applied, nitrogen not used by the crop may be leached into the groundwater, runoff into surface water, or be released from the soil as nitrous oxide. Additional nitrogen in groundwater and surface water has been linked to health effects in humans and affects aquatic ecosystems. Studies have also considered a wider range of adjustments in tillage, grain drying and other field operations.

— *Tillage.* Minimum and reduced tillage technologies in combination with planting of cover crops and green manure crops offers substantial possibilities to reverse existing soil organic matter, soil erosion, and nutrient loss and to combat potential further losses due to climate change. Reduced and minimum tillage techniques have spread widely in some countries but are more limited in other regions. There is considerable current interest in transferring these techniques to other regions.

— *Improved short-term climate prediction.* Linking agricultural management to seasonal climate predictions (currently largely based on the El Niño Southern Oscillation Phenomenon), where such predictions can be made with reliability, can allow management to adapt incrementally to climate change. Management/climate predictor links are an important and growing part of agricultural extension in both developed and developing countries.

The Socio-economic Capability to Adapt

While identifying many specific technological adaptation options, Smit, concluded that necessary research on their cost and ease of adoption had not yet been conducted.

One measure of the potential for adaptation is to consider the historical record on past speeds of adoption of new technologies. Adoption of new or different technologies depends on many factors: economic incentives; varying resource and climatic conditions; the existence of other technologies (transportation systems and markets); the availability of information; and the remaining economic life of equipment and structures (e.g., dams and water supply systems).

Specific technologies can only provide a successful adaptive response if they are adopted in appropriate situations. A variety of issues has been considered, including land-use planning, watershed management, disaster vulnerability assessment, consideration of port and rail adequacy, trade policy, and the various programmes countries use to encourage or control production, limit food prices, and manage resource inputs to agriculture. For example, studies suggest that current agricultural institutions and policies in the USA may discourage farm management adaptation strategies such as altering crop mix, by supporting prices of crops not well-suited to a changing climate, providing disaster payments when crops fail, or prohibiting imports through import quotas.

Existing gaps between best yields and the average farm yields remain unexplained but many are due in part to socio-economic considerations; this adds considerable uncertainty to estimates of the potential for adaptation particularly in developing countries. For example, Baethgen, found that a better selection of wheat variety combined with improved fertilizer regime could double yields achieved at a site in Uruguay to 6 t/ha under the current climate with current management practices. Under the UKMO climate scenario, yields fell to 5 t/ha, still well above 2.5-3.0 t/ha currently achieved by farmers in the area. On the other hand, Singh, concluded that the normal need to plan for storms and extreme weather events in Pacific island nations creates significant resiliency. Whether technologies meet the self-described needs of peasant farmers is critical in their adaptation. Other studies document how individuals cope with environmental disasters, identifying how strongly political, economic and ethnic factors interact to facilitate or prevent coping in cases ranging from the Dust Bowl disaster in the USA to floods in Bangladesh to famines in the Sudan, Ethiopia and Mozambique. These considerations indicate the need for local capability to develop and evaluate potential adaptations that fit changing conditions. Important strategies for improving the ability of agriculture to respond to diverse demands and pressures, drawn from past efforts to transfer technology and provide assistance for agricultural development, include:

1. Improved training and general education of populations dependent on agriculture, particularly in countries where education of rural workers is currently limited. Agronomic experts can provide guidance on possible strategies and technologies that may be effective. Farmers must evaluate and compare these options to find those appropriate to their needs and the circumstances of their farm.
2. Identification of the present vulnerabilities of agricultural systems, causes of resource degradation, and existing systems that are resilient and sustainable. Strategies that are effective in dealing with current climate variability and resource degradation are also likely to increase resilience and adaptability to future climate change.
3. Agricultural research centres and experiment stations can examine the 'robustness' of present farming systems (i.e. their resilience to extremes of heat, cold, frost, water shortage, pest damage and other factors) and also test the robustness of new farming strategies as they are developed to meet changes in climate, technology, prices, costs and other factors.
4. Interactive communication that brings research results to farmers and farmers' problems, perspectives and successes to researchers is an essential part of the agricultural research system.
5. Agricultural research provides a foundation for adaptation. Genetic variability for most major crops is wide relative to projected climate change. Preservation and effective use of this genetic material would provide the basis for new variety development. Continually changing climate is likely to increase the value of networks of experiment stations that can share genetic material and research results.
6. Food programmes and other social security programmes would provide insurance against local supply changes. International famine and hunger programmes need to be considered with respect to their adequacy.
7. Transportation, distribution and market integration provide the infrastructure to supply food during crop shortfalls that might be induced in some regions because of climate variability or worsening of agricultural conditions.
8. Existing policies may limit efficient response to climate change. Changes in policies such as crop subsidy schemes, land tenure systems, water pricing and allocation, and international trade barriers could increase the adaptive capability of agriculture.

Many of the above strategies will be beneficial regardless of how or whether climate changes. Goals and objectives among countries and farmers vary considerably. Current climate conditions and likely future climates also vary. Building the capability to detect change and evaluate possible responses is fundamental to successful adaptation. Thus, even without having clear predictions of climate change, is it possible to identify some strategies that reduce potential vulnerability.

Importance of Feedbacks and Linkages

During the last decades concern about negative impacts of changing atmospheric conditions has increased. Initially most of this concern was focused on local and regional air pollution, but later global aspects were also considered. One of the major global concerns is the increase of 'greenhouse gases' (GHGs: H_2O, CO_2, CH_4, CO, CFCs, N_2O, tropospheric O_3 and its precursors) that influence the earth's radiative balance. The atmospheric concentrations of these gases have significantly increased since the start of the industrial revolution. Such increase could well lead to large changes in regional and global climate.

The sources and sinks of the different GHGs are heterogeneous and difficult to evaluate comprehensively on a global scale. The concentrations of these gases are influenced by a series of chemical processes, many of which are intrinsically linked to each other. Besides these important chemical processes in the atmosphere, the final concentrations are strongly determined by several oceanic and terrestrial processes. For example, oceans take up atmospheric CO_2 through diffusive and biological processes and the magnitude of both is determined by climatic properties, the oceanic energy balance and the ocean circulation. Processes altering the terrestrial sources and sinks are important determinants of many different GHG fluxes between the atmosphere and biosphere. Sources, sinks and fluxes of GHGs from the terrestrial biosphere are more difficult to quantify, because they are defined by local soil, vegetation and climatic properties and the actual land use. All these factors strongly differ geographically and interact with each other on specific time scales. Changes in these fluxes are currently dominated by human activities. This means that any comprehensive assessment of future atmospheric levels of GHGs requires incorporation of these heterogenous spatial and temporal properties, and of human activities affecting the terrestrial biosphere.

Physical and Biogeochemical Feedbacks

Feedback processes influence the exchange rates of GHGs between the atmosphere, biosphere and oceans and modify the residence time of those gases in the different compartments. Feedback processes largely determine the dynamic response of the earth system. These processes include not only the natural geophysical feedbacks, such as changes in climate and ocean circulation, and biogeochemical feedbacks, such as changes in biological activity, atmospheric chemistry and oceanic CO_2 uptake, but also the important anthropogenic feedbacks such as changes in human activities, energy use and land use.

Geophysical feedbacks alter the radiative forcing characteristics of the atmosphere and include changes in characteristics and distribution of clouds, ice and snow. These feedbacks are incorporated in the global climate models and their strength determines the sensitivity of the climate system to changes in radiative forcing. Biogeochemical feedback processes involve changes in the sources and sinks of greenhouse gases and changes in surface properties, such as albedo and transpiration. Direct and indirect effects are distinguished.

Direct effects are those processes that influence atmospheric and ocean chemistry and biospheric processes. The most important feedback process is probably the enhancement of plant growth by increased levels of atmospheric CO_2 This so-called CO_2 fertilization effect has been observed in many greenhouse and ecosystem experiments. Increased CO_2 levels also improve the water-use efficiency (WUE) of plants and change plant growth under water-limited conditions.

The indirect biogeochemical feedbacks involve the consequences of climate change. A large number of those effects have been identified. Examples are changes in ocean circulation, shifts of vegetation zones, and changes in plant growth rates. A temperature increase influences photosynthesis and respiration in a complex manner, enhancing plant growth in regions with temperature constraints. However, enhanced plant growth does not automatically lead to an increase in C storage within an ecosystem. Litter decay is simultaneously stimulated.

Finally climate change alters, like an increased WUE, the distribution of vegetation and agricultural crops, Climate strongly defines the large-scale vegetation and crop distribution patterns and this correlation has be used to assess the global impacts of climate change.

Human impacts and other (e.g., successional) influences are not taken into account. The effects of changes in WUE are simulated by shifting the drought tolerances in the BIOME model. The maps clearly show the large regional differences. Vegetation shifts due to a warmer climate can mainly be observed in the boreal zone, while the effects of increased WUE are mainly located in the tropics, where forests and savanna ecosystems expand.

Importance of Land-Use Change

Sustaining an increasing population requires a continuous flow of agricultural and forest products, such as food, fodder, timber and fuelwood. These products all require land to be produced. Increasing this flow requires more productivity. Productivity can increase through intensification of agricultural practices, improved cropping systems and increase in agricultural areas. Such changes influence the biogeochemical properties of land and are therefore important determinants of the size of sources, sinks and fluxes of GHGs. Up to now only the obvious and large-scale conversions of land (deforestation, reclamation of wetlands, etc.) have been addressed with respect to their importance for global C fluxes to the atmosphere.

Tropical deforestation resulted in a large flux of CO_2 and other GHGs. Forests are felled, burned and removed to create pasture and croplands. These land-use practices are common under swidden agricultural systems. During recent decades, however, deforestation rates have been accelerating and the resulting flux of CO_2 accounted for 25% of the total increase of atmospheric CO_2. After a while large parts of the reclaimed land were abandoned, often degraded, and secondary forest developed. Recently more attention has been paid to the balance between deforestation and forestation in the tropics. This balance is important with respect to the total global fluxes. The resulting heterogenic spatial and temporal patterns make a precise assessment of the total flux from deforestation difficult.

Many evaluations of the total 'land-use' flux are calculated through the total global C budget. Data are used from the observed atmospheric concentrations, the assumed uptake by the oceans and the well-defined emissions from fossil-fuel combustion. The global budget is then balanced with a derived flux from the terrestrial biosphere. The magnitude of the terrestrial flux needed from terrestrial biosphere is generally smaller than the deduced emissions from tropical deforestation. The conclusion of such a deconvolution exercise is that a significant sink is needed to balance the global C cycle. This missing sink is assumed to be located in the terrestrial biosphere, because the processes determining all other fluxes appear to be well understood. In recent years many studies claimed to have identified at least part of this missing sink, but the actual location and the processes involved are difficult to localize and observe. The missing sink emphasizes the importance of land use, land cover and land management in assessing sources, sinks and fluxes of GHGs.

The above discussion has mainly focused on C, but similar conclusions can be drawn for other greenhouse gases, such as CH_4 and N_2O. For example, changes in land cover alter the uptake of CH_4 by soils, different agricultural practices lead to changed CH_4emissions, N_2O emissions are influenced by the timing and amount of fertilizer applications. Such examples indicate that the spatial pattern of GHG emissions from the terrestrial biosphere is very heterogeneous and influenced by physical, biogeochemical, socio-economic and technical factors. Actual land use and its resulting land cover are important controls. Especially when mitigation policies are evaluated, globally aggregated assessments are no longer valid. State-of-the-art assessments should be dynamic, geographical and regionally explicit and include the most important aspects of the physical subsystem, the biogeochemical subsystem and land use and changes therein.

Model Approaches

Many approaches to model the consequences of global climate change have been developed. Most of these models only cover one or a few aspects of the whole chain of emissions, processes and impacts. Emission models largely focus on the energy sector and are therefore often derived from macro-economic or technology models. Emissions from land use are not adequately covered in such emission models. More process-oriented models simulate aspects of atmospheric chemistry, climate and the biogeochemical cycles. The most advanced models are General Circulation Models (GCMs) that simulate the temporal and spatial characteristics of the climate system, and C-cycle models, that simulate the global C budget. The most diverse group of models addresses impacts. Many different sectors have their own model approaches which all simulate one or more aspects of changing atmospheric composition and/or climate change. In particular, agricultural, ecological and hydrological impact models have been developed. Efforts are now under way to harmonize the large diversity in impact assessment approaches and probably an improved, standardized methodology will be agreed upon.

Recently an approach has emerged that integrates different sources of GHG emissions with the processes in the atmosphere and, if evaluated as necessary, with different impacts. Among the first of such models was IMAGE (Integrated Model to Assess the Greenhouse Effect). This model simulated the dynamics of different GHG emission sources and computed dynamically the final atmospheric GHG concentrations considering the influence of atmospheric chemistry and the C cycle. These concentrations were fed into a radiative forcing model that computed a globally average annual climatic change. The only impact considered was sea-level rise. The model was applied to develop the first global IPCC scenarios.

The innovative approach of this IMAGE model has now several competitors. The successor of IMAGE was ESCAPE, an integrated model that was developed for policy evaluation in the EU. The structure of IMAGE has been improved by including different impact modules in a geographically explicit way. The ESCAPE framework has been the basis for two further model developments. Hulme *et al.*, concluded that the complete ESCAPE framework was too complex and developed a derived, highly parameterized global model that is suitable to determine the sensitivity of the climate system and rapidly evaluate the consequences of different policy options. However, feedbacks and linkages between subsystems are not adequately covered in this global model.

Currently several groups are involved in developing such integrated assessment models with a more adequate treatment of feedbacks and linkages, further integrating the structure of the ESCAPE framework. Some of them focus on only a single region, such as the Asian-Pacific region or North America, but include high-resolution impact modules and linkages between them, while the trends of other regions in GHG emissions are prescribed and the global biogeochemical and physical processes are simulated in a more aggregated manner. Other global models are being developed by several institutes. One of the few fully implemented and documented integrated global models is IMAGE 2. Its innovation includes a dynamic simulation of land cover change, driven by socio-economic, atmospheric and climatic factors. The IMAGE 2 framework is especially developed to link all relevant processes comprehensively, while simultaneously retaining regional and local characteristics.

The Image 2 Approach

The overall objective of the IMAGE 2 model is to simulate, on the basis of political and socio-economic scenarios, plausible future trends of GHG concentrations in the atmosphere and to determine their impacts on physical, biogeochemical and human systems. To accomplish this, important scientific and policy aspects of global climate change must be linked. The time horizon of the simulations should be several decades in order to be able unambiguously to discriminate the effectiveness of policy measures. This objective leads to several scientific goals, such as to provide insights into the relative importance of different linkages and feedbacks in the earth system, to provide estimates of sources of uncertainty and to help identify gaps in our knowledge and data in order to help set the agenda for climate change research. The related policy-oriented goals are to provide a quantitative basis for analysing the societal cost and benefits of various measures to address climate change.

Here only the purpose of each IMAGE submodel is summarized. For a full description of all models and their linkages, reference is made to the papers in Alcamo,. The IMAGE model consists of three fully linked submodels: the Energy-Industry System (EIS), the Terrestrial Environment System (TES) and the Atmosphere-Ocean System (AOS). The input data of the socio-economic models are mainly prescribed economic, technologic and demographic trends combined with different control policies, while the environmental models are mainly driven by climatic, vegetation, crop and soil properties.

The models making up EIS compute the emissions of GHGs in 13 larger, somewhat homogeneous regions as a function of energy consumption and industrial production. End-use energy consumption is computed from various economic driving forces. Sector-specific emission factors are obtained from literature and calibrated against data for global and regional emissions. The models are designed such that the effectiveness of improved energy efficiency and technological development on future emissions in each regions can be evaluated and that the set of models can be used to assess the consequences of different policies and socio-economic trends on future emissions.

TES simulates the changes in global land cover based on climatic and socio-economic factors. The roles of land cover and management are used to compute the fluxes of CO_2 and other greenhouse gases from the terrestrial biosphere to the atmosphere. The structure of TES is discussed in more detail below.

The AOS computes the build-up of GHGs in the atmosphere and the resulting climate change averaged over latitudinal bands. AOS thus computes transient changes in climate resulting from changes in all GHG emissions. As a starting point the atmospheric concentration of CO_2 is altered through CO_2 uptake by the oceans. Then, temporal trends of the average tropospheric concentrations of GHGs are computed accounting for the chemical atmospheric reactions involving O_3, OH and CO with other gases and sulphate aerosols. The final atmospheric levels of the GHGs (including HO) are used to compute the earth's energy balance. The cooling effect of backscattering of solar radiation on sulphate aerosols is included. Surface heat exchange with land and oceans, and changes in albedo from land cover change, are considered in the final calculation of temperature change. Other climatic characteristics (e.g., precipitation) are crudely coupled by scaling the latitudinal temperature change towards the results from GCMs.

TES and EIS are linked to each other by the demand for biofuels (fuelwood and biomass). This linkage allows IMAGE 2 to evaluate the land cover consequences for future demands for biofuels. EIS and TES are linked in several ways with AOS. The most important linkages are through the emissions of GHGs and changed albedo of land cover patterns. AOS provides EIS and TES with a changed climate that is directly used to determine the energy demands for heating and cooling, compute potential vegetation patterns, crop distribution and yields, and modify the rates of N_2O emissions. The final concentrations of CO_2 also directly influence plant productivity. These dynamic linkages allow IMAGE 2 to be directly applied to assess the sensitivity of the system for different feedback processes.

All submodels have their specific domain with their own spatial and temporal resolution, scale and dimensionality. Despite this heterogeneity, we have tried to integrate all submodels through their obvious linkages. All socio-economic models (EIS and the agricultural demand and land-use models) distinguish 13 major regions. These models are thus regionally explicit and can be calibrated and validated with a large series of different tabular databases for countries and regions. All environmental models for the terrestrial biosphere are implemented on a 0.5x0.5° grid. Each grid cell is characterized by its climate, soil, and land cover. Spatial heterogeneity in environmental characteristics and local environmental influences on biogeochemical processes are thus adequately covered. Innovative rule-based systems are used to integrate all these dimensions comprehensively.

Modelling Land Cover Change

Modelling land cover change is difficult. Although many models simulate some effect of land-use change, such changes involve often only prescribed tropical deforestation. Contraction, expansion and shifts of agricultural lands, intensification, and different crops and forest management practices are rarely included. Such changes have to be included in global emission models because their cumulative effects are significant. We have developed a land cover change model that is forced by a heterogeneous set of physical, biogeochemical and socio-economic variables. Although, many other models and studies stress the importance of macro-economic driving forces, we have not emphasized those.

Our approach is based on a set of well-established ecological and agricultural models. Potential vegetation patterns are modelled with the BIOME model. This model uses climatic variables to delimit the distribution of different Plant Functional Types (PFTs, cf. deciduous broad-leaved trees, evergreen needle-leaved trees, C_4 grasses). Each PFT is characterized by its specific climatic constraints defined by the accumulated heat during the growing season, drought, and frost tolerances. Biomes are assemblies of these PFTs. BIOME allows us to evaluate shifts in vegetation due to climatic change and changes in water-use efficiency and the consequences of these shifts on total C sequestration.

Crop distributions are modelled using the agro-ecological zone approach. Crops require a minimum growing period (the period with adequate moisture and heat supply) to grow and mature. This growing period, together with several other climatic parameters, defines the potential rainfed distribution of crops. Yield is computed with a simple model that distinguishes different crop types (C_3 or C_4, tropical or temperate, legumes or non-legumes) and estimates biomass production for each crop under the prevailing climatic conditions during the growing period. The economic yield is a fraction of this production and is adjusted for soil quality. The result of these two models is thus a potential vegetation and crop yield for each 0.5x0.5° gridcell.

Transformation of land cover is governed by the local, regional and global demand for agricultural commodities (grain, meat, wood and biofuels). This demand is satisfied by specific land use in croplands, pasture lands, rangelands and (managed) forests. Land use is strongly related to the potential productivities and technically achievable yields for each gridcell. We developed an agricultural demand model to estimate the societal demand for agricultural products. These calculations are performed for each world region and use the same socio-economic input as EIS.

Regional agricultural demand is calculated from the per caput human consumption for different crops and meat products based on a assumed elasticity relationship between consumption and per caput income. These elasticity coefficients are estimated from 1970-1990 data of FAO,, and are the main adjustable parameters in this demand model. Total human (as opposed to animal) demands for these products are then computed for a given income and population scenario. A similar procedure is used to compute the total regional meat demand. This demand is then converted into numbers of livestock, and their feed requirements (concentrates from crops and residues and roughage from range and pasture lands). These feed requirements are added to the total regional human crop demand. This demand is then adjusted for trade between regions. The roughage demand is converted into a demand for grassland by multiplying productivity per animal and the average required rangelands per animal.

The total regional demand for agricultural products is reconciled with the potential local distribution and yields of crops, computed earlier. Land cover is initialized for 1970 conditions using an aggregated version of the database of Olson *et al.*,. The agricultural crop demands are assumed to be satisfied in the single class 'agricultural land', while the demand for rangelands can be satisfied by several types of grasslands, depending on what occurs locally. The basic idea of the dynamic land cover change model is to change land cover until the total regional demands are satisfied. The model thus generates future land cover logically consistent with demand and production potential. This is probably not realistic but it is congruous with available global data sets and some basic driving forces. Improvement of the analyses, incorporating different land-use systems, political structures and cultural characteristics, is far beyond the current requirements and analysis in IMAGE 2. However, we are eager to learn from a better understanding from emerging projects and experiences with other models.

Actual changes of future land cover patterns are difficult to estimate, because the actual driving forces are heterogeneous, differ per region and are in general poorly understood. Our approach is therefore to prescribe a set of transparent logical rules to match regional land demand with its potential. These rules, together with a management factor, define future patterns. For example, there are two weighted rules that guide the assignment of new agricultural areas: (1) new areas should occur adjacent, if possible, to existing agricultural land, because of the availability of infrastructure, transport and population; (2) the new land should develop in those areas of the highest production potential. Similar sets of rules are specified to satisfy rangelands and fuelwood demand, but the demand for cropland is satisfied first. Land not under agricultural management is assumed to succeed towards the potential vegetation as determined by the BIOME model. This land cover model results in a dynamically changing land cover on the terrestrial grid. The land cover patterns are consistent with a changed climate and are used to drive the terrestrial C cycle. One major improvement of this approach over earlier models is that an increase of agricultural area does *not per se* lead to deforestation, but it is a function of the agro-ecological suitability of all types of land. Agriculture could thus expand into grasslands as well as forests.

Carbon Fluxes

The C-cycle model estimate the C sources, sinks and fluxes resulting from natural processes and land cover change. The C-cycle model computes, for each land cover type, Net Ecosystem Productivity (NEP), which is the difference between the Net Primary Productivity (NPP) and soil respiration. NPP is a characteristic of each land cover type, but its actual value is corrected according to local (temperature and moisture availability) and global (atmospheric CO_2 concentration) conditions and the time passed since the last land cover change. NPP and soil respiration are modified under the influence of climate change, assuming characteristic response functions for photosynthesis, plant respiration and soil respiration. CO_2 fertilization is a function of available PFTs in each land cover type (such asC_3 C_4, annual, perennials or trees), temperature, moisture, soil fertility and altitude. The simulated NEP is consistent with the simulated climatic conditions.

NPP is partitioned into land cover specific biomass components (leaves, branches, stems and roots), each with a specific longevity, and non-living matter (litter, humus and charcoal), each with a specific turnover time. We have initialized the model starting from 1900. Most IMAGE 2 simulations start therefore (in 1990) with NEP conditions that are close to equilibrium with prevailing climatic and atmospheric conditions. This near-equilibrium condition shift when the environment changes or land is converted into another type. If natural vegetation is converted, a fraction of the biomass is burned on site, immediately releasing CO_2, while the remainder is allocated to the non-living pools. Such change results in an additional long-term low CO_2 flux from these sites. If agricultural land is abandoned, we assume that the potential natural vegetation returns. However, such succession takes time and we assume that NPP slowly builds up following a logistic function.

Land-use Emissions

The Land Use Emissions model relates global land use to the flux of emissions of CH_4, N_2O, CO_2 NO_x, and Volatile Organic Carbons (VOCs, excluding CH_4). In addition, the model estimates the emissions resulting from biotic processes unrelated to human activity, such as N_2O emissions from soils in unmanaged forests, and trace gas emissions from aquatic systems. The calculations are important in determining land use and cover related GHG emissions and can be used to evaluate strategies for reducing these emissions.

This model is based on current estimates for the various sources and species. Given the limitations of the available data, the model presents grid-based estimates for a number of different sources, including CH_4 emission from rice, wetlands, emissions of CH_4, CO, NO_x, N_2O and VOC from deforestation, savanna burning and agricultural waste burning and N_2O from natural soils, N fertilizer and deforestation. For some of the sources (landfills, domestic sewage treatment, termites, CH_4 hydrates and aquatic sources) geographically explicit calculations are not yet possible because of data limitations.

Applications of Image 2

The IMAGE 2 model has been tested and calibrated against data from 1900 to 1970 and can for this period reproduce trends in regional energy consumption and energy related emissions, land cover, terrestrial flux of CO_2 and emissions of other GHGs into the atmosphere. The time horizon for the actual simulations extends from 1970 to 2100 and each simulation is characterized by a specific set of input options that describe policy and socio-economic trends. Here we present a scenario which is an adaptation of the intermediate IPCC scenario. The scenario makes 'conventional' assumptions about future demographic, economic and technological driving forces. There are no climate-related policies, except those that were already agreed upon (e.g., Montreal Treaty on CFC emissions). This scenario further assumes an increase in the energy use efficiency and a regionally specific vehicle utilization. This baseline scenario can be used as a baseline against which to compare other scenarios.

Land-use change and emissions

We compared the emissions calculated for the years 1970, 1990 and 2050 for CH_4 and N_2O with different reference estimates for selected land-use-related. From these reports either the methodology has been adopted, giving by definition similar results, or emission factors have been adapted to arrive at equal global totals for 1990. In some cases our results slightly differ from those of the reference studies, for example for CH_4 emissions from animal waste. Major differences are found for all emissions from deforestation. The global deforestation rate for the baseline scenario is 155 000 km^2/year for the period 1970 to 1980 and 140 000 km^2/year for 1980 to 1990. Official inventories of deforestation by FAO, show that rates may have been respectively 115 000 km^2/year and 155 000 km^2/year. Reference estimates for biomass burning are based on these FAO deforestation estimates for the early 1980s and result therefore in lower emissions than our baseline scenario.

Our calculations are sensitive to assumptions of demographic, technological and economic development and the productivity of land (crop yields and animal productivity). Slight changes in the balance between agricultural demand and production could cause large differences in land use and associated emissions. There is a disagreement between the baseline scenario and the data from Watson *et al.*, for future agricultural emissions, as shown for rice. The assumed increase in rice yield (t/ha) in the baseline scenario for South and East Asia and the Centrally Planned Asian countries is 1.1% per year for the period 1990-2010, or a 25% increase over these 20 years. The assumed annual yield increases by Watson *et al.*, are respectively 1.2% and 0.5%. Both estimates are far less optimistic than those of FAO, that assume a global rice yield increase of 1.5% per year for the same period leading to a total yield increase of about 35%. This illustrates the uncertainty in the forecasts of rice productivity and area and the consequences for the associated CH_4 emissions. In general, our baseline also assumes far lower animal productivities than FAO,.

The increase in crop productivity is coupled to increases in the use of agricultural inputs, represented in the model by N fertilizer and a management factor. The annual increase in global N-fertilizer use in our baseline scenario is about 2% for the period 1990-2025 and 0.3% for 2025 to 2050. These assumptions lead to a doubling of N-fertilizer use between 1990 and 2025. The associated N_2O losses are directly related to fertilizer use. It is not unrealistic to assume that fertilizer use will continue to increase along with the projected increases in crop productivity. However, the projected increase in both the baseline and Watson *et al.*, for the coming decades is somewhat slower than for the period 1970-1990.

The baseline scenario generates globally a large increase in the extent of agricultural land. Although the global population more than doubles, the increase in agricultural area is only around 15%. This is comparable to the observed increase during the last half century. This relatively small increase in extent is accompanied by a large increase of agricultural productivity on existing agricultural land.

Although the global increase is modest, there are large regional differences. Agriculturally highly developed regions (especially North America and Europe) increase their productivity significantly, while the population is largely stable. Part of the increase is exported, but there is still a large amount of land that is abandoned. We assume that this land returns slowly to its potential vegetation (often forest), which creates a significant C sink in these regions. The decrease in agricultural land starts immediately in the simulation. Latin America follows a similar path during the next century, but initially there are still significant amount of deforestation.

Other regions (especially Africa, India and China) display different patterns; the increase in population and income (and its linkage to changed dietary preferences) amplify agricultural demand. This demand cannot be satisfied by an increase in productivity and imports and leads to a large increase in agricultural land. By the end of next century almost all suitable agricultural land will be occupied. This will result in deforestation and consequently large GHG fluxes from these regions. The model thus simulates a shift of the major loci of deforestation from Latin America towards Africa and Southeast Asia. The strongly increased human pressures on land in these regions could have severe consequences for the development of sustainable land management, food security and biodiversity.

Forestation as a mitigation option

The model allows not only the evaluation of the consequences of different socio-economic trends, but can also be used to evaluate different mitigation options. There are many different mitigation options, but their effectiveness or suitability can differ per region. It is therefore required to develop a comprehensive capability to evaluate such options. IMAGE 2 provides some possibilities, from which we present two examples.

One of the most often presented mitigation options is forestation. This option increases the C sequestered in forests and remains effective if the total extent of forests continues to increase (NEP of old-growth forests are in balance with respect to the C cycle), or if the resulting forest products are used in a durable way (construction wood), replace other CO_2 sources (woody biomass could replace fossil fuel) or both (construction wood could replace cement).

The simulation of IMAGE 2 satisfies the demand for agricultural products first. This demand drives land cover patterns and the remainder is natural vegetation with its specific C dynamics and storage. Only those regions that climatologically can be forests are suitable for the forestation mitigation option, but a large portion of these regions is used for other purposes. Early estimates of forestation potential were therefore probably much too optimistic because, to offset large amounts of fossil-fuel-related CO_2 emissions, large areas have to be forested annually. Forestation in shelterbelts, along fields and roads and in degraded lands is, although important for other reasons, alone not sufficient.

Biofuels as a mitigation option

Another mitigation option is the use of biomass for energy generation. An efficient use of renewable biofuels and biomass for generation of energy could decrease the dependence on fossil fuels. Such a programme is already being implemented in Brazil, where ethanol from sugar cane is used to replace petrol. Biofuels applied for transportation are derived from oil crops or from fermentation of sugar or starch. The techniques involved are not yet competitive with energy derived from fossil oils. However, new technologies could make very efficient use of biomass as an energy source. For example, the foreseen development of fuel cells, in which biomass is converted into a hydrogen-rich gas, promises the possibility of small- and large-scale electricity production with an efficiency ranging from 50 to 80% with few environmental impacts. Many future scenarios assume therefore large amounts of energy derived from renewable biomass. Especially in the second half of the next century when fuel-cell systems should be widely operational, biomass could be the major source of renewable energy.

We make several assumptions for the use of biomass as an energy source in the baseline scenario. The use of such biomass increases from 5 EJ now (10^{18} J: current global energy use is 300 EJ) to 75 EJ in 2050 and 210 EJ in 2100. These figures are based on reasonable mixes of energy carriers, the annual growth in energy demand induced by the increasing global population and its income, a decline in oil and natural gas resources and the preference for renewable energy sources over fossil fuels. We assume for the baseline that the required biomass is simply available from crop and forest residues and waste. This amount of biomass thus does not require new cropland. However, the biofuels used in the transportation sector (ethanol) are still derived from sugar cane and maize but this relatively small additional demand has been included in the simulated extent of agricultural land.

We derived two different scenarios from the baseline scenario. These scenarios only differ with respect to their source of biomass-energy level. In the second scenario, biomass crops, we assumed that such large amounts of biomass are not simply available and that a large portion should be grown specifically in the form of short-rotation forests and other biomass crops, such as *Miscanthes* spp. These crops are effective sources, because they generate a high yield of 40-60 t/ha/year and have a wide potential distribution. We assumed that 40% of the total biomass-related energy demand is satisfied by such crops. The third scenario, no-biomass fuel, is the opposite approach. Renewable energy from biomass is not feasible and the total modem biomass fuels are replace by fossil oils.

The differences in the simulated extent of agricultural lands are large. The large amount of additional demand for biomass crops result in a 30 or 65% increase in, respectively, 2050 and 2100. This increase leads to both a larger deforestation and expansion into more arid regions, such as steppes and savannas. Here the importance for a comprehensive evaluation with respect to land use becomes apparent. The increasing land requirements in the biofuel-crop scenario result in a smaller total sink in the terrestrial biosphere. This has consequences for the simulated atmospheric CO_2 concentration and is the main reason that this scenario yields higher concentrations than the baseline, where it was assumed that biomass was freely available. The consequences of biomass energy scenarios must be carefully evaluated and land-use issues should not be neglected in such assessment.

The added value of integrated assessment models is thus their ability to generate insights that cannot be easily derived from individual natural or social science component models that have been developed in the past. The current integrated assessment models with respect to global climate change all have a mix of policy and scientific goals. This has lead to a large diversity of different, but complementary approaches. Only a few models include a simulation of land use and land cover change. This omission limits a comprehensive evaluation of feedbacks and effective linkages among economic, energy, agriculture and forestry sectors and withholds adequate evaluation of many C cycle or land cover related mitigation options. The above examples illustrate the usefulness of these integrated models in creating a greater understanding of the linkages and relationships among the climate system, the biosphere and human activities. Models such as IMAGE 2 and AIM will be enhanced continuously. More consistent scenario developments, better algorithms and better initialization and validation data will improve future applications. A major conclusion from such analysis should be that from the integration of a series of disciplinary expertise and insights emerges a better understanding, which could further disciplinary and multidisciplinary research.

Without such understanding, integrated global change assessments will continue to emphasize energy-industry-transport emissions, a globally aggregated C cycle with limited land-use change, and the climate system, while natural ecosystems and agrosystems will only be addressed in impact studies. Excluding them from integrated assessments of climate change will lead to non-sustainable climate policies. Fortunately, major international research programmes (Diversitas, WCRP, IGBP and HDP) are now addressing land use and land cover change more systematically, which means that we now enter an exciting new era in global change research.

Climatic Change and Agriculture: Status and Trends

The predicted changes in climate, especially increased atmospheric CO_2, temperature and precipitation, associated with changes in nitrogen deposition, tropo- and stratospheric ozone levels, UV-B radiation, etc. can have great impacts on world agricultural production and supply patterns. In order for agricultural production to be sufficient to meet the demands of the ever-growing human population, the impact of the climate must be understood and integrated in any future planning. The Food and Agriculture Organization of the United Nations is much concerned with this issue. The Organization formed an Interdepartmental Working Group on Climate Change and charged it with coordinating FAO activities in the critical area.

A Consultation 'Global climatic change and agricultural production: direct effects of hydrological and plant physiological processes' was held from 7 to 10 December 1993 at FAO Headquarters in Rome, with the support of the United Nations Environment Programme in Nairobi.

The objectives of the Consultation were: (1) to analyse and assess the effects of higher atmospheric carbon dioxide (CO_2) levels, higher ultraviolet (UV-B) radiation, higher near-surface ozone concentrations, higher temperatures and changing precipitation/evapotranspiration ratios on plant growth and food production; (2) to provide an overview of the state of knowledge on individual and combined effects, including a description of the processes, and availability of data for specific crops; (3) to identify gaps in our present knowledge and to focus on critical research needs.

The meeting was attended by 40 external participants, the 12 members of FAO's Climate Change Group and other interested FAO staff. The discussions concentrated on crops, with limited consideration of natural ecosystems. Because the majority of the presentations and discussions emphasized agricultural systems, it was decided to focus the book on agriculture alone. However, we believe that there are strong interactions between agricultural and natural ecosystems and these natural ecosystems play a significant role in the global carbon cycle which can greatly impact agriculture. The two are separated for convenience only. No discussions took place on the effects of climate change on oceanic biotic production because of the absence of specialists in that field among the assembled groups.

Due to increasing consumption of fossil fuels such as oil, gas and coal, in order to satisfy human needs, large quantities of carbon dioxide are being emitted into the atmosphere. In addition, agricultural and industrial activities add considerable amounts of methane (CH_4), nitrous oxides (N_2O) and chlorofluorocarbons (CFCs) to the atmosphere as well. Collectively, all of these gases lead to what is called the enlarged greenhouse effect because all of them absorb infrared radiation. It is estimated that about 50% of CO_2 emitted to the atmosphere remains in it and the other 50% is taken up by the ocean and terrestrial ecosystems.

General circulation models based on the equivalent of doubling CO_2 concentration have predicted a global increase in mean global temperature ranging from 1.5 to 4.5°C. Furthermore, all current models show that the increase will be unequally distributed globally. For example, it is predicted that temperature rise in higher latitudes will be much more than in equatorial regions. These models also predict a change in precipitation with some regions receiving more rain than the present and others receiving much less rain.

Agriculture is totally dependent on weather and climate. Despite much effort by climatologists, there is considerable uncertainty about the potential impact of climate change on this sector. Little is known as to how, when, where and to what extent climate change will occur; one incontestable fact is the rising concentration of carbon dioxide (CO_2) in the earth's atmosphere. An additional certainty is the soundness of the basic greenhouse theory: the composition of the gas mix in the atmosphere strongly affects the planet's temperature. If the model scenarios are realistic, correctly reflecting future realities, such an increase may have serious consequences for agriculture and, in particular, for the regional food security in some regions. This security is already strained by increased demand and intensification of resource use, by the fast-growing human population and by an increase in per caput consumption of agricultural products. There is little question that agriculture must keep pace with the burgeoning human population which is expected to reach 10 thousand million in the 21st century. Since the timing, spatial pattern and magnitude of climate change are uncertain, policy-makers face a dilemma as to what measures, if any, should be taken to face the predictions of climate change. At present, some of the proposed preventive and mitigation measures would have enormous economic and social costs, particularly in relation to energy use. But these costs could even be larger in the future, if uncertainties are not resolved and we would learn that changes in the climate would be larger and more devastating than initially thought.

Thus, agriculture faces a particularly difficult dilemma: should it begin adapting at high cost to uncertain climatic changes while seeking the resolution of the scientific issues concerning the magnitude of climate change and its impact on agriculture, or choose the 'business-as-usual' principle and run the risk of leaving future generations unprepared when changes materialize? Although clear-cut answers might not be available for at least the next decade, improving scientific knowledge on the agronomic and ecological effects of any climate change, both adverse and positive, and on the ability of humans and ecosystems to adapt, might reduce the uncertainty and help formulate better policy. Therefore, whatever policies are adopted should be subject to frequent reassessment and must be flexible enough to accommodate change dictated by the gaining of new knowledge.

The panel of experts assembled by FAO, made up of persons with a wide range of backgrounds in various aspects of agriculture and natural ecosystem science, reached consensus on the following issues.

Effects on Plant Physiology and Soil Processes

CO_2 fertilization and anti-transpiration effects

As plant type and plant productivity are major determinants of food production, it is critical to understand and quantify the response of the most important crops to changing environmental conditions. Different crop species have different responses to increased atmospheric CO_2concentrations and to the combined changes in other factors, such as temperature, precipitation, pollutants, ultraviolet radiation (UV-B), etc. For example, free-air CO_2 enrichment experiments (FACE) in Arizona, USA, show that cotton is highly responsive to elevated CO_2whereas wheat is much less responsive.

CO_2 is a key factor in photosynthesis and in plant growth. After diffusion into the plant through stomata, it is transformed by photosynthesis into carbohydrates. A large number of water molecules are lost by transpiration through thc stomata for cvery CO_2 molecule entering the leaf. In a CO_2-rich environment, the larger concentration gradient forces more CO_2 into the plant, while partial closure of the stomata will reduce water losses from the leaf. As stomatal opening decreases in a high CO_2 environment, water loss from the plant is also reduced, increasing water-use efficiency. For example, there are indications that doubling present CO_2 concentrations reduces stomatal conductance (opening) by 30-60%

depending on the species. The reduction in water consumption is called the CO_2 anti-transpiration effect. The water-use efficiency (WUE) of the plant improves, since less water is used for equal or more CO_2 transformed into dry matter. At the same time, net photosynthesis might increase, because photorespiration, which reduces carbon gain, is less at high CO_2 concentrations. In optimal conditions of light, moisture and availability of nutrients, this fertilization effect could increase above- and below-ground biomass production by 10 to 40% depending on crop type and even to higher levels such as the case in cotton. Earlier and more rapid leaf production is expected, and the incremental increase in biomass can benefit the root system even more. The earlier establishment of ground cover, because of the early canopy development, may also limit water loss by direct soil evaporation. This response is even greater at high temperatures, since the optimum temperature for photosynthesis increases with high atmospheric CO_2. In contrast, combined low temperatures and elevated CO_2 concentration could reduce plant growth.

The importance of the anti-transpiration and fertilization effects varies with crop type. For example, at double atmospheric CO_2 the biomass production of C_3, plants, including major crops such as rice, wheat, potatoes, beans, soybean, sunflower, groundnut and cotton, can be expected to increase, on average, by some 30%, provided other factors are not limiting.

On the other hand, and independently of CO_2, the physiology of C_4 plants permits a generally higher photosynthetic capacity than in C_3plants. This efficiency, however, is quickly saturated at increased CO_2 concentration. Therefore, in a CO_2-rich environment, the net improvement in photosynthesis of C_4 plants is proportionally small (about 10%, mostly in stem), although the WUE might significantly improve (by about 40%). This category includes crops of major importance for food production, such as maize, sorghum, sugar cane and millet, but also tropical grasses, pasture, forage and some weed species that are critical for agricultural production. CAM plants seem to be less sensitive to CO_2 enrichment as well.

Effects on soil fertility

A sudden doubling of the atmospheric CO_2 concentration and associated higher temperatures - as in most experimental set-ups, both enclosed and under free-air enrichment - may result in soil degradation including nutrient depletion. A potential increase in initial soil fertility under elevated atmospheric CO_2 can be expected if the increase in CO_2 occurs gradually, as in practice and as is the case in crop models that take into account the gradual transient increases. Additional litter is likely to raise soil organic matter content unless litter chemistry drastically changes which would cause a decline in litter decomposition rate. However, the higher soil temperature can stimulate microbial respiration and the decomposition of the organic matter (i.e., mineralization) and cause the release of nutrients which become available for plant uptake through the root system unless microbial competition with plants for the available nutrients is intensified. Furthermore, since both the nutrient uptake efficiency and the structure (length and density) of root systems improve under elevated CO_2 concentrations, overall plant nutrient uptake can also increase. Furthermore, there are indications that the expanded root system can penetrate more deeply into the soil and reach extra sources of moisture and nutrients.

An additional driving force is an enhancement of the symbiotic association between root systems and the fungi and bacteria of rhizosphere under high CO_2 concentrations. If these concentrations do not greatly alter the existing gas composition in the rhizosphere, the colonization of the root system by mycorrhizae (fungi in symbiosis with roots, facilitating plant uptake of occluded phosphate) and nitrogen-fixing bacteria (bacteria with the ability to incorporate atmospheric nitrogen into nitrogenous compounds which can be utilized by living organisms) is likely to improve the nutrient uptake by the host plants. However, it is possible that this process will ultimately slow down if changes in litter chemistry, e.g., an increase in C/N ratios and an increase in tannins, occur. In such a case there will be a greater demand for the use of additional fertilizers.

Uncertainties

This promising picture of improved food production under higher atmospheric CO_2 is modified by other factors that can limit growth. First, most of our understanding of the positive effects on crops relies on short-term and controlled studies usually at the individual plant level. Despite some evidence from field experiments such as FACE, extrapolation and generalization to large-scale field conditions or to long-term global food production are still uncertain. Also, since crop responses to climate change are site-specific and species-dependent, the knowledge on one kind of grouping or plants (e.g., annuals) may have little relevance to other species or groupings (e.g., perennials). Second, under conditions of limited soil nutrients or solar radiation (e.g., through enhanced cloud cover), higher CO_2 may not improve overall yields; in much of the world such stress conditions are the rule rather than the exception. In fact, nutrients and other present-day limitations might be more critical to agricultural production than the potential impact of climate change especially if urbanization and increased demand for agricultural products push agriculture toward marginal land.

Experimentation and scientific explanation provide additional arguments for a balanced assessment of the CO_2 fertilization and anti-transpiration effects. There are feedback mechanisms that might lower the direct effects of higher CO_2 and temperatures. For example, because transpiration rate per unit leaf area goes down when stomatal conductance decreases, leaf temperature can rise, triggering a general increase in canopy temperature and potentially higher water use. There are indications that the foliage ages more rapidly and the critical period of seed filling is shortened at high canopy temperature. Such feedbacks can undercut the decrease in crop seasonal water use that might result from the improvement in water-use efficiency of plants.

There are physiological and biochemical mechanisms that can further limit the long-term benefits of CO_2. For example, after an initial improvement, the growth rate of many perennial species exposed to elevated CO_2 concentrations tends to fall to near that of unexposed plants. One reason is a possible biochemical feedback, such as the reduction in the enzyme rubisco due to the accumulation of carbohydrates in leaves, that slows down photosynthesis and reduces phosphorus availability to carry carbohydrates into the growing parts of plants. There is also insufficient knowledge of the exact causes and consequences of this downregulation in various crops. It appears that sink strength plays a major role in this process as downregulation seems to occur more strongly in crops which are sink limited than in crops which are sink unlimited. The direct physiological responses of plants (e.g., changed stomatal resistance, leaf area index) may be different at different growth stages to varying CO_2 concentrations above and beyond the finding that early stages of plant development are more responsive to elevated CO_2 relative to later stages.

Interactions with pollutants in the atmosphere are also important for agriculture. Tropospheric ozone (O_3) levels, sulphur dioxide (SO_2) concentrations and UV-B radiation are likely to increase in parallel with the rise in CO_2 Depending on the species, it is estimated that crop damage from peak O_3 exposure may result in significant reduction in yield in the proximity of the source (e.g., downwind of major urban areas). Because most studies are performed in closed systems, there is considerable uncertainty on the applicability of such results in open-field environments where significant spatial and temporal variability of O_3 concentration are to be expected. Moreover, the damage attributed to O_3 alone is confounded by the interaction among elevated CO_2, increase in temperature and changes in water availability. For example, under some conditions, yield losses attributable to water stress can outweigh those resulting from 0. exposure. Yet, all indications suggest that the increase in tropospheric O_3 will have a negative effect on crops; what is unknown is the magnitude of this effect at the local scale or in combination with other factors.

Losses of stratospheric ozone are contributing to higher levels of UV-B radiation but, although some experimental work suggests reduced growth, there is little field evidence that this would significantly affect plant growth. Furthermore, it seems uncertain that the interaction between UV-B and ozone will be of importance to crop productivity.

Effects on Hydrological Processes

Many countries allocate a large percentage of their freshwater resource to agriculture. Since water is a critical factor for crop yields, it is obvious that the advantages of elevated atmospheric CO_2 cannot be realized if water is limiting. There is strong evidence, however, that responsiveness to CO_2 increases in dry years in some grasslands. Although some recent models show increased drought in some regions and increased rainfall in other regions, available information on the potential effect of climate change on the global water availability is conflicting and remains largely fragmented, except for the nearly gratuitous statement that the hydrological cycle and ultimately the overall water supply are likely to be affected.

Current models predict that there may be an increase in the Indian monsoon and a decrease in moisture availability in the Amazon region and parts of tropical Africa. Numerous Global Circulation Models (GCMs) have attempted to predict these changes in rainfall patterns based on increased global warming potential. In most cases the predictions compare poorly with the observed data. There is no solid evidence that recorded changes in precipitation in the last three decades are related to the increasing atmospheric CO_2. Three reasons are cited for the mismatch between real and simulated data obtained from GCMs. Firstly, precipitation is the least reliable model output, because of the temporal and spatial variability in rainfall and more importantly because physical processes governing short-term variations are not well understood. Secondly, runoff resulting from precipitation depends on a large number of parameters many of which are specific to each river basin. GCMs rarely account for these parameters; their predictions are therefore inaccurate as long as the horizontal processes (i.e., overland flow, channel routing, groundwater movement, etc.) governing local water balance are largely ignored. Finally, many GCMs generally do not take into account specific cloud effects on regional heat balance, the influence of oceans, or extreme events (e.g., drought, flooding, etc.), the frequency of which is predicted to increase with climatic change.

However, in spite of their inherent deficiencies, models are valuable analytical tools for the assessment of the potential impacts of climate change on the water resources systems. Often these models use hydrological time series (e.g., measurement of river flow) to associate past meteorological events (e.g., precipitation) with recorded changes in lake levels and river flow regimes. This is not justified in view of the non-linear relationship between rainfall and runoff; small changes in precipitation have a magnified effect on river flow. For example, when regional precipitation decreased ~20% around Lake Chad, flow in the local river system was reduced ~50%. Similarly, in the Volga basin where regional precipitation is already increasing, the discharges will substantially increase if GCM predictions are correct. In regions with significant changes in precipitation, monitoring of river flow might therefore be a good indicator of any 'climate change effect'. However, even when based on this indicator, the analysis of hydrological time series of major lakes and river systems (Lake Chad, River Nile, Caspian Sea, Great Lakes, etc.) fail so far to detect any trend likely to be caused by global warming. Factors other than climate change (e.g., population growth, industrial and agricultural development, new methods of irrigation) seem to have a far greater impact on water resources.

Effects on Global Food Security

The important question remains: how do the above processes influence the global food production system? Models have attempted to simulate climate change impacts on the world food situation by pulling together population data, components of the earth-atmosphere-ocean system and scenarios of climate change as well as estimates of potential changes in yields of important crops. The purpose of these models is to predict changes in land productivity and the geographical shift in agricultural land use as a function of changes in climate and food demand. The main outputs of these models reveal some new predictions, and confirm previous ones, that a doubling of atmospheric CO_2 is likely to cause the following:

1. agro-ecological zones would shift because of temperature increase and improved water-use efficiency, with significant regional differences;
2. crop yields and winter grazing in mid- and high-latitude regions (i.e., mostly developed countries) would improve because of increased photosynthesis, longer growing periods and extension of frost-free growing regions, provided optimum growth conditions are maintained, e.g., by judicious fertilizer and biocide use on agricultural land;
3. in most developing countries, crop productivity would diminish (some 10% reduction in cereals), which could raise agricultural prices on local and world markets and increase the need for cereal imports, although the global food supply/demand ratio might change only little;
4. there can be much risk in tropical and subtropical regions, and the greatest risk to food security would be in Sub-Saharan Africa. The magnitude of the threat will also depend on the behaviour of non-agricultural sectors of the economy in the future.

There remains much uncertainty around these predictions. For example, soil conditions in part of the new lands becoming available through shifting climatic zones may be unsuitable for sustainable crop production. New crop varieties may have to be developed. Also, the predictions are confounded by the uncertainties about the role of volcanoes, oceans and terrestrial ecosystems in global carbon fluxes. For example, recent evidence from the Arctic tundra suggests that due to soil warming the system is already a net source for atmospheric carbon. Present and future alterations of land use (deforestation, extensification or intensification of land use, etc.) are extremely important determinants of the terrestrial carbon fluxes into the atmosphere. At the same time, since elevated CO_2 concentrations can increase net primary biomass production, some enhancement of the terrestrial sink (i.e., more carbon storage in vegetation and soils), is to be expected, at least in the short term.

The inherent weaknesses of the current GCMs further exacerbate these uncertainties. In spite of recent advances, only few models, the so-called Coupled General Circulation Models (CGCMs) consider the role of oceans and the more realistic transient scenario (i.e., progressive build-up of CO_2). Spatial and temporal resolutions are still poor at the regional level and of little use at the local level where farmers must act. Important parameters, such as cloudiness, atmospheric aerosols, ocean-atmosphere-terrestrial linkages, cost of technological and social adaptations, extreme climatic events, lateral water transfers, availability of water resources and soil nutrients, and multiple and inter-cropping systems must be included in the models.

Implications and Needs

Our current scientific knowledge provides a good understanding of plant physiology, morphology and growth. For example, plant response to varying temperatures, changing precipitation, soil moisture and humidity are well documented. Likewise, there are some vegetation models that adequately correlate the present distribution pattern of vegetation and climate. Nevertheless, a high degree of uncertainty still clouds the potential impacts of climate change on both managed and natural ecosystems. The reasons for uncertainty have been discussed in the previous sections. They relate mainly to the paucity of relevant climatic data, the incomplete understanding of the processes underlying the global circulation models, the coarseness of their spatial resolutions, and the insufficient knowledge on the long-term, direct and indirect biological and physical responses of crops and other ecosystems to elevated CO_2 and other correlated factors. The working groups at the Expert Consultation identified several research areas which would reduce uncertainty, improve knowledge, increase preparedness in the face of climate change and provide better grounds for policies related to climate change:

1. Firm quantitative assessment is needed of site-specific crop responses as a function of time and growth stage, in particular for perennial crops and crops of greatest importance to food production. Such assessment needs to be properly structured, and its parameters should be identified and tested for priority crops and regions. Open-field multi-variable experiments should consider important plant biological stresses along with changing atmospheric composition (CO_2, O_3, SO_2, N_2O and other non-CO_2, greenhouse gases), precipitation, and UV-B radiation.

2. As some of the soils of the new lands made available through shifting of climatic zones may be unfavourable for crop production, it is important that they are categorized and mapped to avoid or lessen the chances of inappropriate land-use choices.
3. The long-term effects of elevated CO_2 on the uptake of nutrients (e.g., nitrogen, phosphorus, etc.) are unknown and studies on this subject should start soon. Nitrogen balances should be linked to plant-soil-water balances, and the impact of elevated CO_2 on biological nitrogen fixation and phosphorus uptake need to be quantified for various crops under field conditions.
4. Agronomists need to work closely with climatologists at a regional level to provide a sound basis for optimizing crop, soil and water management under the changing conditions. Special attention should be given to evaluating probabilities of extreme events (droughts, floods) and their effects on plant growth and yield. This should be accompanied by concerted projects on plant breeding by conventional techniques and by using genetic engineering and selection for stress-resistant genotypes.
5. The known physical mechanisms regulating plant water use should form the basis for assessing water balances for drought- or flood-prone regions, in specific soil conditions, and for selected crops and changing climatic scenarios. Water management assumes even a greater importance under a changing climate. Water resources harnessing criteria (e.g., for irrigation structures) should achieve the necessary flexibility to accommodate future changes, particularly in vulnerable areas of the world. Long-term, continuous climate and hydrographic observations are essential not only for detecting climate change signals but also for judicious water resources development.
6. A concerted effort should be directed to increase the quality of global modelling projections. High-resolution databases are needed on land use and land cover, soil carbon, nutrients and mineralization rates, crop cultivars and their climatic requirements, N_2O emission rates from fertilizer applications related to local climate, soil and vegetation conditions. Remote sensing, field-level evaluation of soil and terrain conditions, monitoring of nutrient inputs and use are important tools for these tasks.

7. Improving high-resolution databases will not only help generate better GCM projections, it will also contribute to the planning of overall development process in the developing countries. Better information improves knowledge and helps devise good policies and sound agricultural management practices. These in turn would increase the resilience of production systems to inter- and intra-seasonal climatic variations and to global climate change. Forums for communicating progress on data collection and availability should be strengthened.
8. Better socio-economic data on household income and expenditure are also critically important. Much of this information does not exist or is not at a suitable resolution and consistency, and therefore is not directly comparable between or within regions.
9. There is a lack of basic biological knowledge about how tree species and forest ecosystems are affected by climate change. The rate at which a particular species is expected to move into a newly available area needs to be studied, as well as the effects of elevated CO_2, alone and in conjunction with other global change parameters on successional, mature, degraded and recovering forests in different world biomes.
10. Readily available assessments should extend their perspectives by enhancing their treatment of indicators of sustainability, with implications of climate change and consequences for greenhouse gas emissions.

References

Boden, T.A., Kaiser, D.P., Sepanski, R.J. and Stoss, F.W. (eds.). 1994. *Trends '93: A Compendium of Data on Global Change.* US Department of Energy, Carbon Dioxide Information Analysis Center, Oak Ridge National Laboratory, Oak Ridge, Tennessee.

Fajer, E.D. and Bazzaz, F.A. 1992. Is carbon dioxide a 'good' greenhouse gas? Effects of increasing carbon dioxide on ecological systems.*Global Environmental Change: Human and Policy Dimensions* 2: 301-310.

Houghton, J.T., Jenkins, G.J. and Ephramus, J.J. (eds.). 1990. *Climate Change: The IPCC Scientific Assessment.* Cambridge University Press, Cambridge.

Rosenzweig, C. and Parry, M.L. 1994. Potential impact of climate change on food supply. *Nature* 367: 133-138.

Ruttan, W. (ed.). 1994. *Agriculture, Environment, Climate and Health: Sustainable Development in the 21st Century.* University of Minnesota Press, Minneapolis.

10

Climate Change Implications for Fisheries and Aquaculture

The threats of climate change to human society and natural ecosystems have been elevated to a top priority since the release of the fourth Assessment Report of the Intergovernmental Panel on Climate Change (IPCC) in 2007. While the importance of fisheries and aquaculture is often understated, the implications of climate change for these sectors and for coastal and riparian communities in general are difficult to ignore. At the same time, fisheries and aquaculture do contribute to greenhouse gas emissions, although in a relatively minor way, and present some opportunities for mitigation efforts.

From local to global levels, fisheries and aquaculture play important roles for food supply, food security and income generation. Some 43.5 million people work directly in the sector, with the great majority in developing countries. Adding those who work in associated processing, marketing, distribution and supply industries, the sector supports nearly 200 million livelihoods. Aquatic foods have high nutritional quality, contributing 20 percent or more of average per capita animal protein intake for more than 1.5 billion people, mostly from developing countries. They are also the most widely traded foodstuffs and are essential components of export earnings for many poorer countries. The sector has particular significance for small island States, who depend on fisheries and aquaculture for at least 50% of their animal protein. Climate change is projected to impact broadly across ecosystems, societies and economies, increasing pressure on all livelihoods and food supplies, including those in the fisheries and aquaculture sector. Food quality will have a more pivotal role as food resources come under greater pressure and the availability and access to fish supplies will become an increasingly critical development issue.

The fisheries sector differs from mainstream agriculture and has distinct interactions and needs with respect to climate change. Capture fisheries has unique features of natural resource harvesting linked with global ecosystem processes. Aquaculture complements and increasingly adds to supply and, though more similar to agriculture in its interactions, has important links with capture fisheries.

*

Climate Change and Fisheries

The oceans are an integral component of the climate system and respond to changes in the system. Natural fluctuations occur in oceanographic conditions – some are annual while others are less frequent, with cycles that may extend over decades rather than years. Still others are longer-term changes, which might occur over thousands of years. Of current concern are the specific impacts of human activity on climate change, which is expected to result in increases in sea surface temperature, global sea level rise, decreases in sea-ice cover and changes in salinity, wave conditions, and ocean circulation. These changes in turn will have an impact on the biological productivity of marine ecosystems.

The impact on fisheries of changes in the biological productivity of marine ecosystems will vary between fisheries and will depend of the specific environmental changes that occur and the particular biological characteristics of each species. Changes in a particular marine environment may become conducive to a rapid growth of a high-priced species found in that environment, while the reverse may be true in other instances. Climate change will also result in modifications of the area of distribution of marine resources. Consequences for the fishing industry could be significant.

An expected characteristic of global climate change is a likely increase in the variability of environmental conditions. Experience already gained in dealing with longer term fluctuations in marine environments, such as those induced by El Niño events, emphasize the need for adaptability. As well, ensuring sustainable economic levels of fishing capacity should be determined with the variability in mind. The effects of climate change on fisheries will impact a sector that is already characterized by full utilization of resources, large overcapacity and conflicts among fishers, and others, vying for alternative uses of marine ecosystems. Thus, climate change adds a further argument for developing effective and flexible fisheries management system in an ecosystem context.

The renewable nature of fisheries and the inherent spatial and temporal variation in stock size are primary characteristics of fisheries. It is the primary goal of fisheries scientists to understand and ultimately predict the timing and extent of this variation. Spatial and temporal variation in stock size are species-specific characteristics and require knowledge of the species life history such as reproduction, growth, etc. Predicting fluctuations in abundance is further complicated because the oceanic milieu is neither static nor homogeneous in its composition. Furthermore, since the ocean and atmosphere are coupled, both the distribution and abundance of fishery resources are capable of being modified by climate.

Climate exerts an as yet incompletely understood linkage with fisheries. The linkage, however, occurs through the transfer of energy from the atmosphere to an ocean surface layer of variable thickness. The depth of this layer varies with latitude, distance from shore, and sea on but is generally less than 100 meters and may be considerably shallower in the presence of a pycnocline, or density gradient. Below 100 meters less than 1 percent of sunlight penetrates even in the clearest ocean water and wind-generated currents, with the exception of western boundary currents like the Kuroshio or Gulf Stream, are seldom important. In the surface layer or euphotic zone, however, several components of climate, including solar radiation, wind and temperature, may impact fisheries. These variables are of course correlated, but their oceanographic implications and hence their impact on fisheries differ.

Solar radiation influences the ocean through heating and evaporation at the air/sea interface and by providing light energy for photosynthesis throughout the surface layer. Biological production in the sea originates with primary production, a photosynthetic process whereby chlorophyll-containing phytoplankton generate organic matter. In this process, phytoplankton take in dissolved phosphorus and nitrogen-containing compounds, nutrients necessary for growth. The phytoplankton are eaten by zooplankton, or by some clupeids, such as sardines, and shellfish.

Zooplankton generally are utilized by small fish, which in turn are consumed by larger ones. The bodies of organisms and faces are then subject to bacterial decomposition or autolysis. The remains, being denser than sea water, slowly sink or are turbulently mixed downward, resulting in a flux of nutrients from the euphotic zone. Unless compensating upward transport of nutrients occurs, the euphotic zone will become impoverished. Turbulent mixing in the arctic and shallow margins of temperate oceans, as well as upwelling, are primarily responsible for the return of nutrients to the surface layer.

Vertical mixing occurs primarily during winter when solar radiation is low, vertical density gradients weak, and winds are strong. Mixing occurs through turbulence and the depth to which it occurs depends on wind speed, direction and duration. A useful index of turbulent mixing in fisheries studies is simply to use the cube of the wind speed which can be employed in exploratory analysis with fisheries parameters through correlation or regression analyses.

Turbulent mixing results in the redistribution of nutrients which, with the vernal warming due to increasing incident radiation, leads to a flowering of phytoplankton termed the spring bloom. The spring bloom is primarily a high latitude process which is followed by the propagation of animals of higher trophic levels. Tropical areas are generally less productive than higher latitudes. This lower productivity is partly explained by the greater incident radiation and hence deeper euphotic zone, which is not mixed with deeper nutrient-rich waters, and partly by lower inputs of continentally derived nutrients via rivers. In addition to turbulent mixing, wind also impacts fisheries through wind-generated surface currents which may result in egg and larval transport away from nursery areas and hence, fluctuations in year-class strength.

Wind-generated surface currents may also control productivity through upwelling. This is a process of water motion whereby cold, deeper waters move upward toward the surface as a result of winds displacing warm surface water. While upwelling may occur anywhere, it is most common along the western coasts of continents. The movement of surface water in response to wind is described by a simple mathematical model and is referred to as Ekman transport. In this model, if the wind blows in the same direction for a sufficient duration, say a few days, it will set a surface layer in motion.

This surface current is approximately 2 percent of the wind speed and to the right of the wind direction in the northern hemisphere (to the left in the southern hemisphere). The movement of this layer sets in motion the layer below it which in turn sets in motion a deeper layer. The speed of each layer is slower than the layer above it and the direction is displaced further to the right (northern hemisphere) or left (southern hemisphere) than the layer above. For modelling egg and larval transport occurring in the upper few meters of the water column, however, it is preferable to use the modified transport model of Stolzenbach et al. in which current speed is 3 percent of the wind speed and 15° off the wind direction. However, Ekman transport involves a deeper water column (basement is usually taken as the depth at which the current speed decreases to 4 percent of the surface speed) and the net transport is 45° to the right or left of the wind direction.

Therefore if the wind is blowing approximately parallel to the shore, the surface water will be displaced seaward and replaced by deeper (down to 200 meters), colder nutrient-rich waters. As a result, upwelled waters are highly productive of fish and other organisms. The most marked coastal upwellings are found off western North America, South America, Africa and Australia. The upwelling along the boundary of the Peruvian Current maintains the large stocks of plankton-feeding fish such as the Peruvian anchovy, Engraulis ringens. Diverging currents also cause upwelling. Where adjacent surface waters flow away from each other, deeper water must also rise. Such divergence and associated upwelling are found around Antarctica, along the Aleutian Island Chain, along the equator, and along the northern margin of the Equatorial Countercurrent.

Impacts of El Niño

One of the most dramatic impacts of climate on fisheries occurs every few years off the Peruvian coast. This phenomenon occurs during the southern hemisphere summer months of December to March, when a transequatorial flow of warm low-salinity water originating from the equatorial countercurrent displaces the northbound Peruvian current. Called El Niño, or Christ Child, because of its arrival during the Christmas season, this intrusion of high-temperature and nutrient-poor water 5-10 degrees further south than normal results in mass moralities of fish, as well as increased evaporation and, consequently, increased precipitation on the adjacent land. Dead fish are washed ashore where they decay; the subsequent formation of H_2S, when combined with sea fog, may even blacken the paint of ships. Sea birds feeding on fish die of starvation and the decline in bird population results in reduced production of guano, the basis of the fertilizer industry.

In 1972, El Niño was very intense; total average production in the euphotic zone during this event was 0.39 g $C/M^2/day$, three times lower than the long-term average off southern Peru. The catch of Peruvian anchovy, which had been 13 million metric tons (19 percent of the world catch in 1970) declined dramatically, while other species increased notably during the following year. In 1972 jack mackerel Trachurus murphyi, until then accounting for under 2 percent of the catch in northern Chilean fisheries, suddenly increased to 10 percent. Dramatic increases since 1972 have also occurred in the catch of the Chilean sardine Sardinops sagax, paralleling the rise of jack mackerel and accompanying the decline in anchovy (Fig. 1). The decline in anchovy catch may be explained by intense exploitation as well as by ecological disruption from El Niño in 1972 73.

An important feature of El Niño is that it may be predictable several months in advance in the larger context of the Southern Oscillation, one of the major irregular, periodic fluctuations of atmospheric circulation. Basin-wide atmospheric and oceanic anomalies appear to be as important as local winds in creating El Niño. Large-scale changes in wind fields far out in the Pacific are of particular interest. The trade winds were unusually strong in 1970 and 1971, but were unusually weak in 1972. The eastward flowing North Equatorial Countercurrent intensified and the South Equatorial Current weakened during the 1972 El Niño. It appears that warm water accumulated in the eastern Pacific, deepening the usually shallow thermocline and covering the colder upwelled waters.

Other examples of ocean/climate phenomena with important implications for fisheries are the variation in western boundary current axes and the formation of eddies or rings from these currents. It has long been observed that the stream axis of currents like the Kuroshio, Gulf Stream and East Australian current meander. These meanders in turn influence the circulation of water on the continental slope and shelf. Since fish stocks may be associated with particular water masses, their distributions will also be modified. If the current meander grows large enough it can separate from the current as either a warm-core or cold-core ring, depending on the temperature of the entrained water relative to that of the current remnant encircling the core.

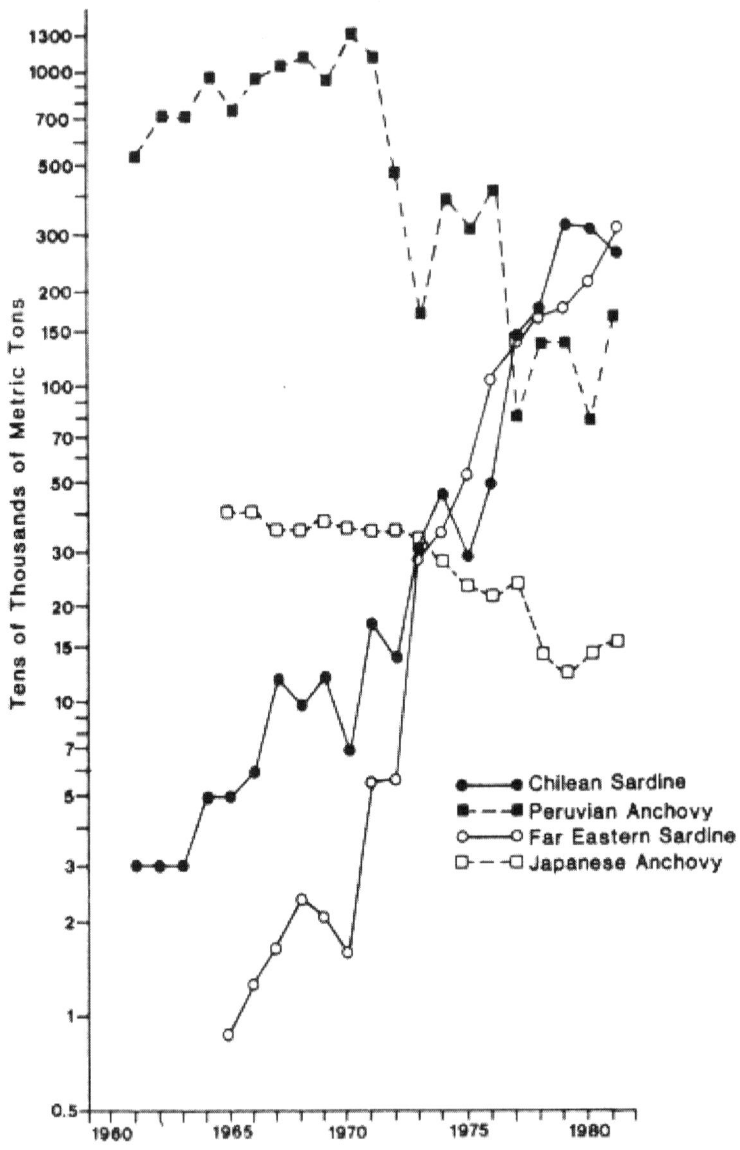

Fig. 1. Interannual variation in catches of sardine and anchovy

The 1982 83 El Niño was perhaps the strongest event of the century and has certainly been the best studied. While analyses are still underway, it is clear that this was not a classical occurrence, but one with a pattern different from earlier events of the century.

Influence of Meanders

Rings that separate seaward of the current rotate cyclonically in the northern hemisphere (anticyclonically in the southern hemisphere); those that separate landward rotate in the opposite direction. These features are 100-300 kilometers in diameter, rotate at several knots, and maintain their integrity for as long as several years. During a ring's life they drift and exchange nutrients, biota and energy with the adjacent water mass. Backus et al. have reviewed the dynamics of cyclonic Gulf Stream rings which are comparable with those of the Kuroshio and with the anticyclonic rings of the East Australian current. These scientists point out that to preserve mass the formation of one ring type (cold- or warm-core) must be in equilibrium with the formation of an opposite type. Based on this assumption and their study of cyclonic rings, they estimate that warm-core rings transfer 3-20 x 10^{15} grams of salt and 10^{21} calories of heat from the Sargasso Sea to the continental slope annually. They further estimate that cold-core rings transfer 5 x 10^{11} grams of carbon as living organic matter from the slope water to the Sargasso Sea, increasing productivity in 10 percent of the Sargasso Sea by 50 percent. The importance of rings to fisheries lies only partly in their role in determining the distribution of oceanic properties. Rings have also been implicated in the destruction of deep water crab pots in the Baltimore Canyon and larger catches of swordfish have been reported from ring boundaries than from adjacent waters.

The cause of meanders and rings is poorly understood; however, Teramoto discusses the Kuroshio meandering as a local manifestation of a large-scale fluctuation in the ocean/atmosphere system. Temperature changes around Oshima Island south of Tokyo may be caused by the meandering of the Kuroshio, and this change is inversely correlated with change in the surface velocity of the Kuroshio. Changes in ocean surface temperature produce changes in heat transmission that are connected with movement in the lower atmosphere. While links between ocean and atmosphere are easily recognized on local or regional scales, there may also be identifiable distant linkages. So-called `teleconnections' exhibit significant relation to variations in widely separated areas of the ocean and atmosphere.

Environmental Change and Stock Fluctuations

There is continuing controversy about the cause of fluctuations in many stocks, since natural (including climatic) causes, or overfishing, or both, can result in a major change in a given fishery. An example illustrating the often equivocal interpretations possible is that surrounding the halibut Hippoglossus stenolepis along the Pacific coast of North America. In the early 1950s it was argued that there was an approximately reciprocal relationship between the catch-per-unit of fishing gear and the number of sets of units of gear.

Fishing effort had been controlled based on this relation, and it was claimed that regulation of fishing effort was responsible for the recovery of catches, which had declined during the 1930s but had increased in the 1940s. Others, however, disagreed. Burkenroad contended that in the decline in abundance of halibut on the Pacific coast west of Cape Spencer between 1915 and 1930, and the increase after that date, both may have been much greater than could be accounted for only by changes in the amount of fishing effort. He suggested that major fluctuations in abundance of this stock might be attributable to natural causes of a regular cyclical sort. Burkenroad thus held that the desirability of applying current theories of biological management to marine fisheries remained to be demonstrated.

Another example is the Far Eastern sardine Sardinops melanosticta. It can be seen in Fig. 2 that the catch of this species is subject to large-scale fluctuation, and this has been recorded since ancient times. The leading causes of the fluctuation have been attributed to both environmental factors and overfishing. The catch of this sardine fell abruptly from a peak of 2,730,00 tons in 1937 to 9000 tons in 1965, a drastic reduction in population size (Fig. 2). Nakai noted that in an area south of the Kil Peninsula a cold water mass had developed during the years 1938-45. He argued that this anomaly of the Kuroshio Current caused mass mortality of the early life stages of this sardine resulting in the depletion of the adult sardine population. In the early 1970s sardine abundance again began to rise and the catch increased rapidly, resulting in a take of 3,610,000 tons in 1981 (Fig. 2).

The fisheries of the Republic of Korea and the USSR were also high. Kondo et al. explained this increase by a combination of environmental factors. In the spring of 1972 the Kuroshio had shifted its pathway along southern Honshu, Japan, from a meandering to a non-meandering pattern, a condition favoring the distribution of copepod nauplii, a small crustacean fed upon by larval sardines. The increased supply of food for larval sardines enabled the 1972 sardine year-class to recover from the earlier stock reductions.

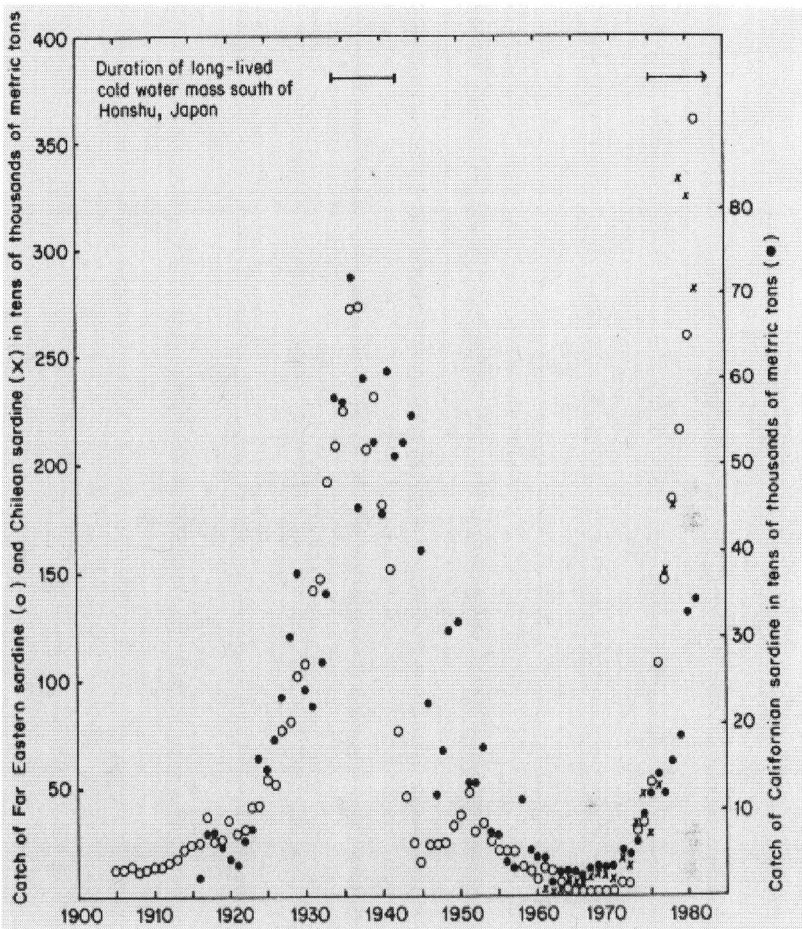

Fig. 2 Large-scale variations in the catch of three sardine species, Far Eastern, Californian and Chilean

While some scientists maintain that the ups and downs of stock level of the Far Eastern sardine have resulted from changes of food conditions of larvae due, to the shift of the Kuroshio, others emphasize the role of human intervention. Cushing points out that in 1930 there were 700 purse seiners for the Japanese sardine, 1000 in 1940. Peak catches occurred in 1936, but subsequently fell dramatically with the failure of the 1938-41 year-classes. The collapse in the early 1940s was followed by a period of decline in which Japanese fishermen concentrated on fish in their first summer of life, and a further decline in catch followed a period of high and sustained effort.

While Nakai associated this failure in recruitment with an anomaly in the flow of the Kuroshio Current in the years 1938-45, Cushing disputes the suggested environmental source of recruitment failure, emphasizing that fishing effort was high on prespawning fish. Murphy also attributes the population collapse to intensive exploitation. He criticizes the environmental hypothesis on the grounds of the apparent lack of a consistent cause/effect mechanism for these years. Environmental arguments have been advanced to explain most fish population collapses, but usually a different causal mechanism has been ascribed to each collapse. Murphy points out that the source of, and difficulty with, these environmental hypotheses is that the ocean climate is highly variable; thus, it is possible to find `significant' shifts or non-shifts in ocean climate associated with almost any biological event one wishes to specify. Murphy concedes that poor year-classes are observable throughout the record, but he contends that they are associated with crashes only when the population is also heavily fished.

Certainly one must be careful in ascribing a change in a stock to a specific oceanic variation. As Murphy shows, it is not difficult to find an oceanic phenomenon which matches a specific change in a stock. Nakai and Kondo et al. explained the fluctuation in abundance of the Far Eastern sardine by shifts of the Kuroshio axis, more specifically, the occurrence and disappearance of a cold eddy south of Honshu. However, when the data are examined closely, it is apparent that it was 1934 when the long-lived cold eddy occurred, whereas stock decline began in 1938.

Although recent recovery of the sardine population was promoted by a strong 1972 year-class, the origin of the recovery existed before 1970. In spite of the frequent occurrence of anomalies of the Kuroshio between 1945 and 1964, corresponding changes of sardine stock were not seen. Advocates of the explanation that fluctuation in stocks lies in fishing intensity also face difficulties. The Far Eastern sardine has been known to fluctuate widely with an irregular cycle of decades to a century, and since ancient times the variation in catch has been associated with the history of success and failure of fishing villages. Such large-scale fluctuations in stock in days past, when fishing effort was very small in comparison with this century, are strong evidence that stock fluctuations can be caused by environmental change.

The solution to this problem does not lie exclusively in the alternatives of environment or fishing. It is an oversimplification to seek the cause of stock changes only in a shift of the Kuroshio axis or solely from the effects of fishing. What is important is that we evolve both theory and methods which consider all sources of variation comprehensively.

Patterns of Fluctuation

Patterns of fluctuation in the number of marine teleosts (all fish except sharks and rays) can be assigned to two broad types.

- Type I: unstable and unpredictable
  - *Subtype IA*: irregular and short-term, e.g. Pacific saury (Cololabis saira)
  - *Subtype IB*: large-scale and cyclical, e.g. sardines (genus Sardinops)
- Type II: stable and predictable, e.g. scombroids (tuna, mackerel, etc.)

The above patterns seem to reflect the characteristics of environmental variation in the sea. Examination of the cycle of production of algae and herbivores may illuminate some of the underlying explanation. Cushing represents production cycles in different regions diagrammatically (Fig. 3). He notes a single peak in the arctic midsummer, a double peak in temperate waters, and only minor oscillations in the tropics. Seasonal differences increase with latitude, as does the delay period between the production of algae and grazing by herbivores.

Reproduced by permission of Cambridge University Press from Cushing Because the tropical cycle is of low amplitude and is continuous throughout the year in a deep euphotic layer, Cushing considers it an efficient cycle. Food is available throughout the year, and both growth and reproduction in the marine animals are probably continuous. In temperate and high-latitude systems, a large quantity of material is transferred from one trophic level to the next at low efficiency, whereas in the quasi-steady-state tropical system less material is transferred more efficiently.

Cushing estimates that the transfer coefficients are three times higher in the `poor' areas than in the `rich' areas. The different cycles correspond to different degrees of stability at higher trophic levels. Cushing notes that a stable community is diverse, that is, there are many species, each few in numbers, such as those found in the deep tropical ocean where the quasi-steady-state cycle of production is found. Because the cycle is of low amplitude, numbers are more or less stable. In contrast, in the high-amplitude cycle of high latitudes and upwelling areas, there are few species, each of high abundance; the community is neither stable, nor diverse. In the case of upwelling, a few types of plankton-feeding fish alternate in dominance with one another at short intervals, a phenomenon known as ‚alternation between species'.

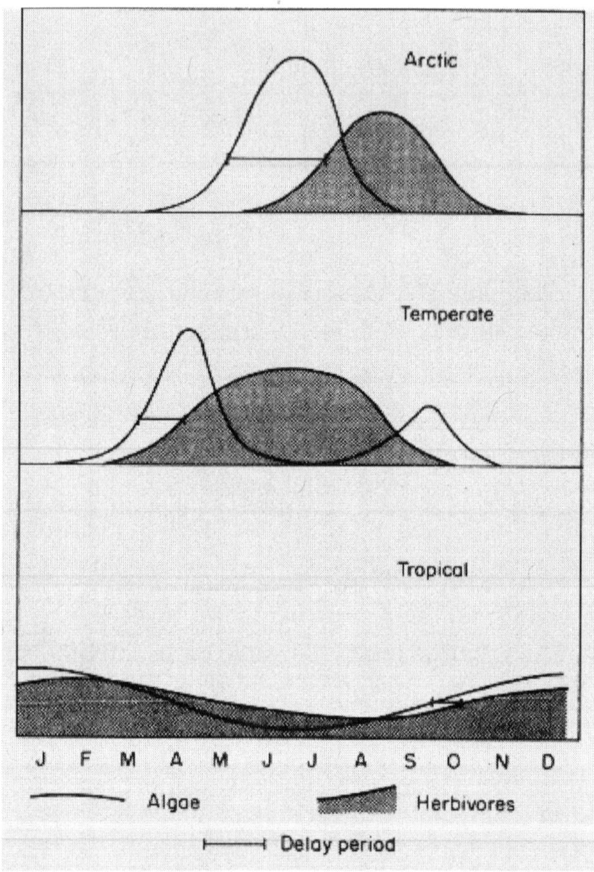

Fig. 3. Diagrammatic representation of production cycles in different latitudinal zones.

Cushing attributes the difference between the inefficient and efficient areas largely to the delay period in the cycle, which depends on the onset of grazing. In the open tropical ocean, production is limited by grazing, possibly at all stages of the cycles, whereas in the upwelling areas and in temperate waters production is controlled by grazing at a later stage in the cycle, by which time a considerable quantity of living material has been produced.

While Cushing's analysis considers the problem of the production cycle in the sea for high latitudes and upwelling areas versus low latitudes, its scope needs to be more comprehensive to reflect adequately the marine environment. In particular, it is necessary to examine the problem from the viewpoint of primary productivity. The productive areas in the sea are the high latitudes (subarctic and temperate), coastal areas, upwelling areas and surface layers. These areas are not only highly productive and rich in biomass, but they are also highly variable.

Food is readily available for animals of higher trophic levels in these areas. In contrast, low latitudes (tropics and subtropics), oceanic areas (in particular the central parts of the oceans), and subsurface and bottom layers are not only lower in productivity and biomass, but are also considerably more stable environments. Obtaining food is laborious for animals in these areas. This situation holds true interannually as well as annually.

Contrasting patterns of fluctuation in number, which differ with life history stages, seem to have evolved in response to these environments. Type I matches the high productivity environment and Type II the lower productivity environment. Type I, in turn, is divided into Subtypes IA and 1B, corresponding to specific temporal scales of environmental variation in the sea. The observed variation in subarctic and temperate seas is considered to result from short-term variation, with intervals of one to a few years, and tends to occur in a relatively small area.

Large-scale variations, with cycles of several decades to centuries, tend to occur in ocean-wide areas. Each may drive the variation in abundance of a variety of species of fish in an area. Fish species of Subtype IA respond more closely to short-term variation in the environment, whereas those of Subtype IB respond more closely to large-scale variation. As shown in Fig. 4, the Pacific saury, Far Eastern sardine, and scombroids, such as chub mackerel Pneumatophorus japonicus and albacore Thunnus alalunga, reveal the fluctuation patterns of Subtypes IA and IB and Type II, respectively.

Interannual Variation

The catch of Pacific saury around Japan has been characterized by shortspaced, interannual fluctuation (as shown in Fig. 4). The fluctuation in catch is thought to reflect a change in stock size, and it is believed that variation of oceanic conditions, especially the pathway of the Kuroshio Extension east of Japan, has caused the fluctuation in stock.

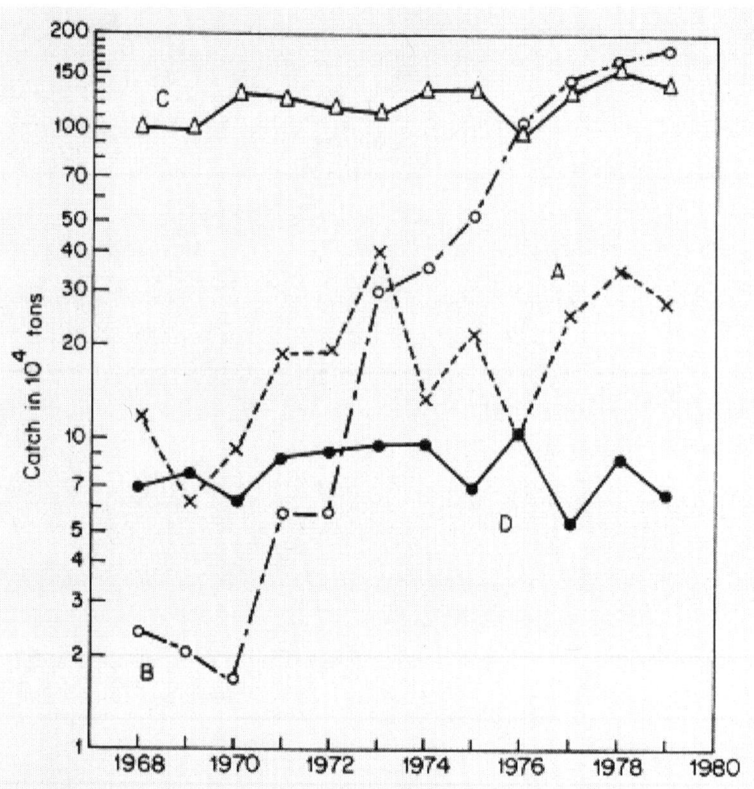

Fig. 4. Interannual variations in the catch of Pacific saury (A), Far Eastern sardine (B), chub mackerel (C), and albacore (D), caught in Japanese fisheries.

Long-term and Phenomenal Variation

This marked phenomenon suggests that variations in stocks of the three species are governed by some common Pacific-wide oceanic variation. A long-lived A-type cold water mass associated with Kuroshio meandering may be regarded as the local manifestation of large-scale variation, such as occurred in 1934-43 and has been occurring since 1975 to the present, a period when sardine stocks have been most abundant (Fig. 2).

Soutar and Isaacs studied temporal change in the scale-deposition rate of the California sardine in cores from anaerobic sediments in the Santa Barbara Basin off northern California. Radiometric methods were used to estimate the age of the laminae, and scale-deposition rates were obtained. Comparison of estimates of the sardine population derived from solution of a fishery catch equation and the scale-deposition rate in the Santa Barbara

Basin indicate a parallel trend. Deposition rates of the sardine scale were very low in 1865-90 and 1940-70, suggesting long-term variation.

Such large-scale fluctuation exhibited by sardines is a function of their ecological characteristics. They are long-lived, fast-growing herbivores living in coastal areas, and these features mean that they are especially subject to changes in oceanic conditions. To summarize, large-scale fluctuation of Sardinops appears to be caused by global-scale environmental variation and depends on whether the sardines are able to utilize a large quantity of phytoplankton or not.

Southern bluefin tuna Thunnus maccoyii is a Type II species distributed over the relatively stable waters of the deep oceans of the southern hemisphere. Fluctuation of stock size is density-dependent, and stock size changes little under unexploited conditions. Fig. 5 shows a relation between the catch by longline and the number of hooks used. The hooked-rate as an index of stock density had been approximately 4 percent from 1954 to 1962, decreasing as fishing intensified, and has fallen to about 0.5 percent since 1975. This is a typical example where heavy fishing resulted in a decline in stock.

Changing Stock-based Interspecies

Changes in stock are brought about not only by environmental and human factors, but by relations between species as well. For example, it has been found that the catch of species of Sardinops (sardines) is inversely correlated with the catch of Engraulis (anchovies). This has been seen in different areas of the world oceans and may be explained by the partly overlapping ecological niches of sardine and anchovy (Fig. 1). The former feeds on both phytoplankton and zooplankton, while the latter depends mainly on zooplankton; both inhabit coastal areas.

Variation Of Fish Species and Fish Biomass

Variation in ocean biota attributable to sustained shifts in ocean climate may be illustrated by the events of the so-called Russell cycle, discovered by Sir Frederick Russell. A northward spread of warm water species occurred in the western English Channel, with warming in the north Atlantic from the 1920s to the 1940s, and was reversed some 30 years later during the early 1970s. The first event was the decline in recruitment of the Plymouth herring stock (Clupea harengus), which began with the 1925 year-class. In 1931, the last recorded year-class entered the fishery, which subsequently collapsed in 1936 or 1937. In the summer of 1926, pilchard eggs Sardina pilchardus were recorded in considerable numbers and remained so until 1960. Between 1925 and 1935, the winter phosphorus concentration declined by one-third. In the autumn of 1931, macroplankton (plankton in the 0.2 2.0 mm size class) declined by a factor of four and the numbers of summer-spawned fish larvae decreased.

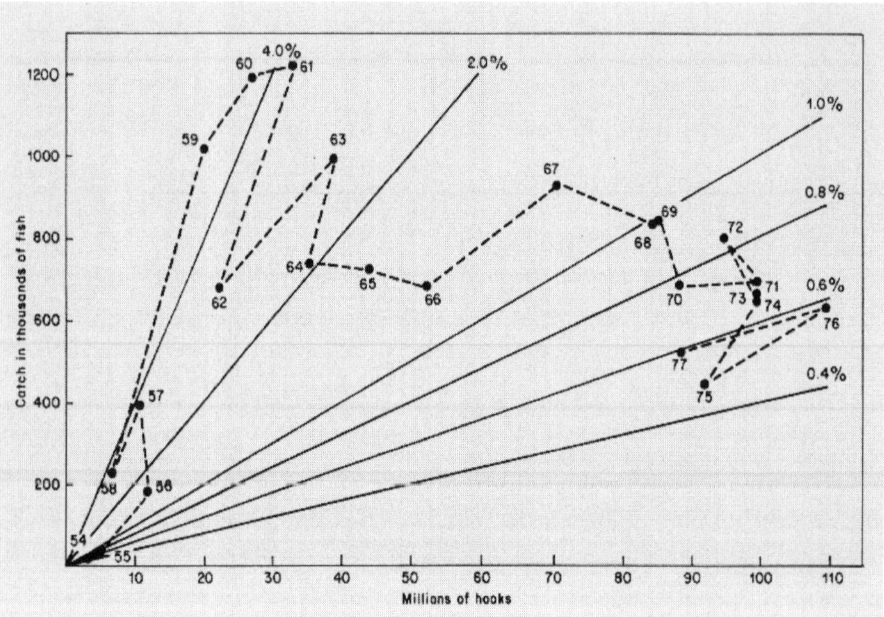

Fig. 5. Relation between fishing effort and catch of southern bluefin tuna in a longline fishery.

When the catches in 1919 22 are compared with those in 1944-52, one of the northern species, ling Molva molva, virtually disappeared, while boarfish Capros aper decreased by an order of magnitude. During the same period, southern species like horse mackerel Trachurus trachurus, hake Merlucius merlucius, red mullet Mullus surmuletus, and red bandfish Cepola rubescens, all increased by an order of magnitude. Phytoplankton in the 1920s resembled those in the 1970s in species composition, whereas in the intermediate years a number of southerly species were found. Furthermore, in 1965, the numbers of spring-spawned fish larvae (generally northerly species) increased again by an order of magnitude. In 1970, the macroplankton increased for the first time since the autumn of 1930, and in the following winter the phosphorus concentration rose to the pre-1930 level. This cycle is summarized in Fig. 6.

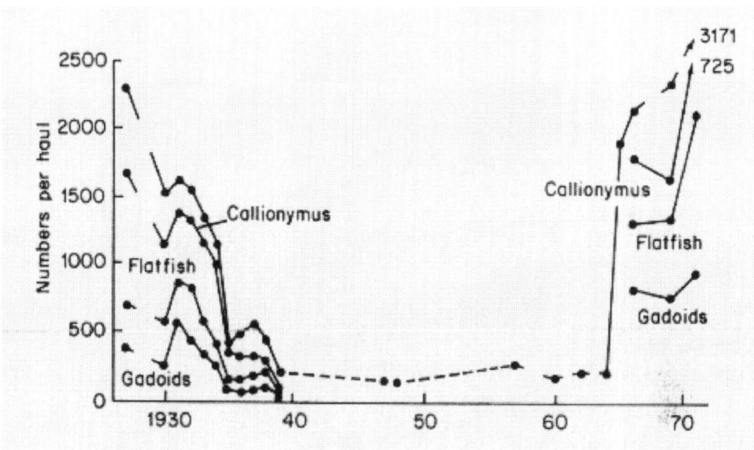

Fig. 6 The Russell cycle in monthly averages of the planktonic stages of teleostean fish, excluding clupeids, between 1924 and 1971 in the vicinity of Plymouth, England.

There are three events common to the two changes in the 1930s and the 1970s. In 1930 macroplankton and winter phosphorus declined, and the spring spawners decreased 5 years later. In 1965 the spring spawners increased in numbers, to be followed by increases in macroplankton and winter phosphorus in 1970. Thus, the later events are the mirror image of the earlier ones; if the decline of spring spawners in 1935 indicated the end of the earlier sequence of events, then their recovery signalled the start of the later ones.

Aquatic Ecosystems and Global Climate Change

Aquatic ecosystems are critical components of the global environment. In addition to being essential contributors to biodiversity and ecological productivity, they also provide a variety of services for human populations, including water for drinking and irrigation, recreational opportunities, and habitat for economically important fisheries. However, aquatic systems have been increasingly threatened, directly and indirectly, by human activities

Freshwater ecosystems and coastal wetlands are essential contributors to the diversity and productivity of the biosphere. Freshwater ecosystems and coastal wetlands are incredibly diverse and productive, providing many tangible and intangible goods and services to human civilisation and welfare (Table 1). Yet these systems are increasingly imperiled by human activity. In the United States, a global center of freshwater biodiversity (Table 2), the extinction rate for freshwater species equals or exceeds that for species in other ecosystem types, including tropical rain forests.

Table 1. Goods and services provided by rivers, streams, lakes, freshwater wetlands, and Coastal Wetlands

	Streams and Rivers	Lakes and Ponds	Freshwater Wetlands	Coastal Wetlands
Water supply				
Drinking, cooking, washing, and other household uses	X	X	—	—
Manufacturing, thermoelectric power generation, and other industrial uses	X	X	—	—
Irrigation of crops, parks, golf courses, etc.	X	X	X	—
Aquaculture	X	X	X	X
Supply of goods other than water				
Fish	X	X	X	X
Waterfowl	X	X	X	X
Clams, mussels, other shellfish, crayfish	X	X	X	X
Timber products	X	—	X	—
Nonextractive benefits				
Biodiversity	X	X	X	X
Flood control	X	X	X	—
Transportation	X	X	—	X
Recreational swimming, boating, etc.	X	X	X	X
Pollution dilution and water quality protection	X	X	X	X
Hydroelectric generation	X	—	—	—
Bird and wildlife habitat	X	X	X	X
Enhanced property values	X	X	X	X
Coastal shore protection	—	—	—	X

Because water flows downhill, aquatic ecosystems occupy low points on the landscape and filter materials that move via gravity from the terrestrial environment. Accordingly, pollutants and fertilisers flowing off the landscape accumulate in these systems, impairing their health. Humans often modify the physical structure of these systems for commerce, agriculture, and recreation, resulting in habitat destruction, one of the major stresses on the integrity of contemporary freshwater ecosystems. Further, these ecosystems are harmed by direct appropriation of freshwater for human consumption, a problem that is likely to grow in the future. Coastal wetlands also receive the pollutants and eroded silts from human activities on land via freshwater inputs. Likewise, they experience direct habitat loss from coastal development and erosion exacerbated by reduced river inflows.

Humans are altering the global climate through the release of greenhouse gases. Average global surface temperatures are projected to increase by 1.5 to 5.8°C by 2100, but increases may be higher in the United States. This projected change in climate will place additional pressure on already-stressed freshwater and coastal ecosystems. Although aquatic systems are generally viewed as resilient and able to maintain a healthy and self-sustaining condition despite large year-to-year variation in hydrologic and temperature conditions, rapid climate change may impose new environmental regimes that will exceed the limits of the resilience of aquatic ecosystems.

Even though aquatic ecosystems have historically experienced elevated temperatures similar to those projected for the next 100 years, the projected rate of change falls outside the natural range of variation and is therefore unprecedented. Moreover, the extent of fragmentation of the landscape in which aquatic ecosystems are situated is unprecedented historically. These systems are increasingly isolated and disconnected, making adjustment to rapid climate change through animal and plant dispersal very problematic. Thus, climate change clearly represents an additional, significant threat to aquatic ecosystems, one that will interact in complex ways with existing human-caused stresses.

Climate Change on Environmental Drivers

The ecological consequences of climate change will largely depend on the rate and magnitude of change in two critical environmental drivers:

1) temperature and,
2) water availability from precipitation and runoff.

These factors regulate many ecological processes in aquatic ecosystems, both dircctly and indircctly.

Average temperatures are predicted to increase across North America, but more markedly at higher latitudes. Some areas of the continent will become wetter and some drier, and the variability in the timing and quantity of precipitation will also change, altering patterns of runoff to aquatic ecosystems. Although the precise geography of these regional shifts is uncertain at present because of limitations in climate forecasting, changes in the fundamental character of many aquatic ecosystems in response to a changing climate are highly probable.

Temperature directly controls many vital life processes, and a change in the thermal regime (e.g., extreme temperatures, their duration, and seasonal rates of temperature change) can directly regulate rates of growth and reproduction for species. Since individual aquatic and wetland species are adapted to a specific range of temperatures, global warming will shift the potential geographic ranges of species to the north, or to higher elevation in mountain regions. Likewise, the more southerly (or lower-elevation) part of the present geographic range of many species will become unsuitable.

Table 2. Global Significance of U.S. Freshwater Species

	Number of Described U.S. Species	*Number of Described Species Worldwide*	*Percentage of Known Species worldwide found in U.S.*	*U.S. Ranking Worldwide in Species Diversity*
Fishes	801	8,400	10	7
Crayfishes	322	525	61	1
Freshwater mussels	300	1,000	30	1
Freshwater snails	600	4,000	15	1
Stoneflies	600	1,550	40	1
Mayflies	590	2,000	30	1
Caddisflies	1,400	10,560	13	1
Dragonflies & damselfies	452	5,756	8	uncertain
Stygobites (cave fauna)	327	2,000	16	1

The availability of species to move into expanded ranges will depend on the availability of habitat and the ability of species to move along dispersal corridors, which will vary for the different types of aquatic ecosystems considered here. The volume of water in an aquatic ecosystem directly influences ecological processes by determining the amount of available habitat and many aspects of water quality. The regional pattern of precipitation and runoff dictates how the volume of water in an aquatic ecosystem changes over time. Because climate naturally varies from region to region (e.g., arid West versus humid East), the hydrologic regime (the seasonal pattern of magnitude, frequency, duration and timing of runoff) also varies among regions.

Seasonal variation in water volume strongly influences what kinds of species can flourish in an aquatic system. Therefore, a change in regional climate that alters the existing hydrologic regime has the potential to greatly modify habitat suitability for many species and cause significant ecological change (even if the thermal regimes remain unchanged).

Streams and Rivers

Stream and river ecosystems comprise both the actual aquatic (in-channel) environment and the associated floodplain or riparian system. The expected impacts of climate change on these ecosystems will depend on how thermal and streamflow regimes deviate from present conditions. The amount of deviation will reflect both the regional and local setting of the ecosystem.

Changes in temperature

An increase in air temperature due to global warming will translate directly into warmer water temperatures for most streams and rivers, thereby altering fundamental ecological processes and species distributions. Streams and rivers are relatively shallow, turbulent, and well-mixed systems, meaning they exchange heat and oxygen easily with the atmosphere. Therefore, they will become warmer under projected climate change.

A warming of water temperatures by, for example, 4°C in present-day ecosystems would represent a northward latitudinal shift in thermal regimes of about 680 kilometers (422 miles), and this would have serious consequences for aquatic ecosystems. The life processes of many aquatic organisms are temperature-dependent. Warmer water can increase growth rates and stimulate ecosystem production.

For example, aquatic invertebrates at the base of the food web (e.g., aquatic insects) may mature more rapidly, albeit to a smaller size, and reproduce more frequently. Assuming no change in food resources, invertebrate production of streams and rivers may increase, potentially yielding more food for fish. However, higher water temperatures will also increase the rate of microbial activity and thus the rate of decomposition of organic material, which may result in less food being available for invertebrates and ultimately fish.

In either case, warmer water holds less dissolved oxygen, so water quality will be reduced for organisms such as invertebrates and fish that have a high oxygen demand. For many species, even modest increases in temperature can have dramatic effects that are not necessarily intuitive. For example, in far northern rivers, many aquatic invertebrates require prolonged periods of near 0°C in winter followed by a rapid rise in spring temperature to cause eggs to hatch. The construction of dams that release slightly warmed water during the winter (i.e., water that does not freeze) has been documented to cause massive local extinctions of invertebrate species for tens of kilometers downstream.

Thus, warmer winter temperatures under global warming could induce similar changes in river diversity and related productivity. Species at the southern limits of their geographic distributions will face local extinction unless they migrate. The ability to migrate to thermally suitable habitat will depend on the dispersal mode of the species and the availability of suitable migration corridors.

Many species may simply move upstream, because headwaters are generally cooler than downstream reaches. However, warmer headwaters could result in the disappearance of species that already are restricted to cool headwaters. So, although upstream movement may work well for some species, an overall reduction in biodiversity in many watersheds can be expected, as cool-adapted headwater species lose critical habitat. This situation could be ameliorated if species lacking thermal refuges within a watershed could migrate northward to more suitable habitat.

In prehistoric periods of climate change, many aquatic species were able to disperse and find thermal refuges. For example, during the late Pleistocene glaciation of about 10,000 years ago, fish moved from north to south in the Mississippi River Basin. Unfortunately, stream and river systems are now highly fragmented owing to human alteration of landscapes (e.g., dams and reservoirs, deforestation, diversion of water for offstream uses such as irrigation and urban development), so migration corridors along many stream channels are restricted.

Some species may avoid this problem by overland dispersal. For example, most aquatic insects have winged adult stages, although some of these species are poor fliers (e.g., mayflies) and would most likely not be successful, particularly in harsher regions such as the arid West. Purely aquatic species like fish and amphibians are expected to have many problems because the overland route is not an option for them, and not all waterways connect to higher latitudes. For example, fish of lowland streams in the desert Southwest and in the southern Great Plains occupy streams and rivers that predominantly drain west and east, providing no opportunity for northward migration.

As many as 20 native fish species in these regions are likely to become extinct under a few degrees of warming. The effect of climate warming on fish has received much attention, because many are highly valued game fish. In one study, Eaton and Scheller estimated increases in maximum weekly water temperatures in streams across the continental United States in response to an average air temperature increase of about 4°C, which is well within the range of projected elevated temperature for the United States over the next century. Their analysis revealed that thermally suitable habitat for 57 species that require cold or cool water (including game fish, such as trout, salmon, and perch) would decline by about 50 percent.

Habitats for many fish requiring warm water were also predicted to decline slightly, although the habitats for some species like largemouth bass and carp, which can flourish in extremely warm water, would increase. Trout and salmon will be particularly vulnerable to warming. For example, in streams of the Rocky Mountain region, an increase in air temperature of only 1°C is predicted to reduce suitable stream habitat for trout by 7-16 percent, and a 3°C increase would reduce habitat by 42-54 percent.

Remnant populations may hold out in locales where local groundwater effluent continues to provide cool-water refuges in warm summer months or thermally suitable hatching sites for eggs. However, over time, even groundwater will warm, and those remnant populations would decline. Ranges for cold-adapted species may expand northward or to higher elevations, where very cold temperatures now limit population growth.

At the northern or high elevation end of current distributional ranges, conditions should improve for trout, because of the favorable warming of very cold surface waters. In Alaska, winter water temperatures are likely to increase by several degrees Celsius, eliminating extensive ice cover and allowing invasion of cool-water species and the range expansion of seasonally migratory species such as salmon and trout.

Commercially important salmon in the Pacific Northwest may also be harmed. During the middle Holocene (6,000 to 9,000 years ago), when temperatures in the Pacific Northwest were 1-2°C warmer and the climate was drier, the very large populations of salmon in some rivers in the region appear to have been depressed. The current depressed populations of Pacific salmonids (due to habitat degradation, barriers to migration, commercial harvest, genetic dilution from hatchery fish, hydrologic alteration, and variable ocean conditions) make them particularly vulnerable to a warmer and drier climate.

For example, warmer temperatures and reduced streamflows in the Columbia River Basin may increase mortality of juvenile coho salmon or create thermal barriers for migration of adult salmon. Juvenile mortality may also increase during periods of warmer spring-summer water temperatures, when warm-water fishes are more efficient predators. Petersen and Kitchell estimate that inter-annual variation in summer water temperatures up to 2°C from 1933-1996 has caused predation rate on juvenile salmon by one predator species to vary by up to 100 percent. Populations of sockeye salmon in Alaska, however, may potentially benefit from a warming of sea-surface temperatures, which favor increased ocean productivity. This pattern has been observed during bouts of natural climatic variation over the last 300 years.

Changes in precipitation and runoff regimes

A modified seasonal pattern of runoff in streams and rivers in response to climate change will alter species composition and ecosystem productivity. The seasonal pattern of precipitation falling on a watershed is translated into surface runoff that feeds into streams and rivers. In ecological terms, the extremes of runoff are critical events that influence species composition and the productivity of aquatic and wetland communities. Streams in the northern tier of states and in much of the mountainous West have snowmelt runoff regimes characterised by a predictable period of high flow in late spring followed by a predictable period of late summer, fall, and winter baseflow. By contrast, in the deserts of the Southwest, winter flows are very variable; summer flows are usually very low because of lack of rainfall, but intense monsoonal thunderstorms can create flash flooding.

In the eastern United States, precipitation is generally high enough to sustain year-round flow in streams. Thus, knowledge of the particular regional changes in streamflow regimes brought about by climate change is critical to anticipating ecological impacts. Some projections are reasonably well supported and others need to be considered as possible scenarios to evaluate the range of ecological concerns related to a changing climate.

Snowmelt runoff

One major consequence of global warming will be a shift from spring peak flows to late winter peaks in snowmelt-dominated regions. Many of the life history characteristics (e.g., reproductive strategies) of both aquatic and riparian species have evolved to avoid or take advantage of the predictable high spring flows in these ecosystems. For example, many fish in snowmelt streams time their reproduction so their young can avoid stressful spring peak flows. Some species lay eggs in the fall that hatch before snowmelt runoff. High winter flows associated with rain or rain-on-snow events can scour the streambed and destroy these eggs.

In streams of the Sierra Nevada in California, native trout that lay eggs in the late spring are favored in years when large winter floods occur, whereas non-native trout that lay eggs in the fall are favored in years when high flows do not occur until the spring, after the young have emerged from the stream gravel. Further, increased streamflow and associated increases in water velocities during winter and spring could reduce the number of salmon successfully returning from the sea because these fish rely on accumulated energy reserves to reach upstream spawning areas.

Cottonwood trees provide important riparian habitat for many terrestrial animals along streams throughout the West, and their ability to successfully reproduce and replace themselves along streams and rivers depends on high spring flows from snowmelt that inundate floodplain habitat. A warmer climate that reduces the magnitude of spring flows could have broad impacts throughout this region. Another significant consequence of a shift from snow to rain in high elevation or northern basins is the reduction in streamflow in late summer. This is expected even if winter precipitation increases in northern latitudes as projected, because excess precipitation will not be stored as snow, which provides a source of runoff to sustain late summer baseflow in arid highlands.

Lower summer baseflows translate into less instream habitat for invertebrates and fish. Further, less water in the stream channel means less water flowing into stream-side groundwater tables, which are important for sustaining riparian tree communities. These ecosystems are likely to experience very significant changes in species composition and productivity.

Growth of precipitation variability

Some climate change models predict possible increases in the intensity of rainfall on fewer rain days, as has already been observed for precipitation data in the United States over the last century. The implications for streams and rivers are significant. For example, in a historical reconstruction of flood histories for upper Mississippi River tributaries over the last 7,000 years, Knox found that small shifts in temperature (1-2°C) and precipitation (10-20 percent) caused sudden changes in flood magnitude and frequency. Because the transport and storage of nutrients and pollutants depend on flow, an increase in floods would probably result in more silt and pollutants entering streams and rivers.

The corresponding degradation in water quality could lead to a loss of sensitive stream species. Further, floods scour the streambed and displace stored organic carbon (food resources for many species), bottom-dwelling organisms, and small fish fry. A well-developed empirical literature suggests that a substantial increase in flood frequency would cause a shift in species composition and perhaps eliminate many species.

Streamflow reduction

Even if flooding increases in magnitude and frequency, earlier snowmelt and higher temperatures could still result in lower summer streamflow in many areas. In addition, some areas could become generally drier. Periods of low flow in streams and rivers are particularly stressful for aquatic and riparian ecosystems. The drying of streams into isolated pools crowds organisms and results in reduced dissolved oxygen levels. In arid regions, periods of prolonged low flows are common, but even in these streams, rewetting is critical to maintaining high diversity.

Prolonged dry periods could selectively eliminate biota that require wetter conditions. In more humid regions characterised by greater annual precipitation, such as the eastern United States, periods of no flow are less common. Accordingly, increasing the duration of no-flow periods in these streams would represent a large deviation from usual conditions and thus be very damaging ecologically. One previous analysis has suggested that a 10 percent decline in annual runoff in eastern streams that have little groundwater inputs could result in nearly half of these small streams ceasing to flow in some years.

Reduced water depths may not only eliminate habitat for aquatic organisms but also increase the vulnerability of sensitive species (e.g., amphibians) to harmful ultraviolet radiation. Many aquatic communities in large rivers are partially dependent on riparian floodplains, either for nursery habitat for fish or for seasonal export of nutrients from floodplain wetlands to the river. If these floodplains become disconnected from the main rivers because of reduced streamflows, aquatic productivity and diversity may decline.

Lakes

Lakes are very vulnerable to climate change. The historical record clearly indicates that during previous periods of climate change, the distribution of lakes changed dramatically as the balance among precipitation, evapotranspiration, and runoff shifted. Many of the physical and chemical features of lake ecosystems depend on the depth of the lake, the amount of heat it absorbs from and releases to the atmosphere, the supply of nutrients to the lake from the watershed, and the retention time of water in the lake.

Together, these factors determine the thermal characteristics and dissolved oxygen availability in the lake, which influence habitat suitability for species and the lake's seasonal and annual ecosystem productivity. Because of differences in size and geographic variations in temperature and watershed runoff among lakes, they provide a very wide range of habitats for freshwater species. Predicting the consequences of climate change for any particular lake is dependent on an understanding of how species composition in that lake will change.

Changes in temperature

Warming of lakes will increase the potential for production of nuisance algae, a phenomenon that will be exacerbated where predatory fish are eliminated due to loss of suitable thermal habitat. As lakes warm in the spring and summer, they develop a less dense upper layer and a cool, more dense lower layer of water. The upper layer has high oxygen levels even into late summer, because winds mix the waters to expose them to the air. Because of high light levels, the surface waters are very productive, especially where watershed nutrient supply is high.

The lower layer is colder and does not mix with the atmosphere. Dead organic material (phytoplankton, zooplankton, etc.) from the productive surface waters falls into these deep waters, where bacteria and other bottom-dwelling organisms consume them and deplete oxygen in the lower depths. The depletion of oxygen means that, by late summer, these deep waters may become marginal habitats for many invertebrates and fish. This condition of oxygen starvation is exacerbated in those lakes that have a high natural or human-caused supply of nutrients.

Indeed, human activity has resulted in significant runoff of nitrogen and phosphorus into most lakes in North America, which are now much more productive (eutrophic) than they would be in the absence of human activities. If large and productive lakes warm too much, they may have insufficient oxygen in deeper, cool water in late summer to support large game fish. Whole-lake experiments have shown that when key predatory fish species are eliminated, smaller fish increase in abundance and reduce zooplankton populations, thereby allowing algae to proliferate. This cascade of effects can directly result in a decrease in water clarity.

Predicting the effects of altered thermal regimes on fish is complex. In many studies, the effects of a doubling of CO_2 on aquatic thermal regimes have been modeled to evaluate responses of individual fish species whose thermal preferences are reasonably well known. In large, deep lakes (e.g., many of the Great Lakes), suitable thermal habitats are expected to increase for almost all fish. Not only will fish thriving in warmer waters benefit, but cold-adapted fish will probably also benefit from a slight warming of deep, cold waters which are expected to retain sufficient oxygen even under a doubling of CO_2 and a temperature increase of 3.5°C.

In smaller and shallower lakes, by contrast, the entire lake volume is likely to warm significantly, and fish requiring cold water (especially large, predatory fish) will suffer from a reduction in habitat. Oxygen concentrations will decline in these warmed waters, further degrading deep-water habitat during the stressful summer months. Even larger lakes, if shallow, could experience a substantial warming and loss of oxygen from the lake bottom, especially if human pollution is substantial, as in Lake Erie.

In very small, eutrophic lakes, by contrast, prolonged ice cover prevents the lake from absorbing oxygen from the atmosphere, and depletion of dissolved oxygen in winter often causes die-backs in fish populations. In these lakes, climate warming may enhance survival of fish in winter. In a thorough study, Stefan et al. simulated the effects of a doubling of CO_2 on 27 lake types (defined by combinations of three categories of depth, area, and nutrient enrichment) across the continental United States and examined the responses of fish species to projected changes in lake temperature and dissolved oxygen. They found that suitable habitat would be reduced by 45 percent for coldwater fish and 30 percent for cool-water fish, relative to historical conditions.

Shallow and medium-depth lakes (maximum depths of 4 meters and 13 meters, respectively) were most affected. Habitat for warm-water fish was projected to increase in all lake types investigated. Because warmer waters support more production of algae, many lakes may become more eutrophic due to increased temperature alone, even if nutrient supply from the watershed remains unchanged. Warm, nutrient-rich waters tend to be dominated by nuisance algae, so water quality will decline in general under climate change.

The possible increase in episodes of intense precipitation projected by some climate change models implies that nutrient loading to lakes from storm-related erosion could increase. Further, if freshwater inflows during the summer season also are reduced, the dissolved nutrients will be retained for a longer time in lakes, effectively resulting in an increase in productivity. These factors will independently and interactively contribute to a likely increase in algal productivity.

Hostetler and Small assessed the sensitivity of lakes across all of North America to climate change using models that simulate seasonal thermal evolution in response to atmospheric forcing. Their models project that under CO_2 doubling, future water temperatures would increase substantially because of increased atmospheric temperatures and reduced winter ice cover. Regional variation in warming would be significant, with surface water temperatures in midwestern and southern lakes and reservoirs increasing by up to 7°C in summer. Many lakes and reservoirs would have surface water temperatures exceeding 30°C, which would be ecologically significant. They found the magnitude of warming to be similar for both deep and shallow lakes within a region.

Watershed runoff changes

Altered stream inflows into lakes can cause complex ecosystem responses in lakes by changing lake depth and modifying water temperatures. One of the most direct effects will be reduced lake levels, although areas that become wetter could have higher lake levels. Even large lakes are vulnerable. Indeed, the extent of potential lake-level declines in some of the Great Lakes is projected to exceed the rate of change projected for sea-level rise.

Permanent lowering of lake levels will expose more shoreline, possibly harming productive littoral (near-shore) zones and coastal wetlands of the Great Lakes. Many of these lakefringing wetlands may become isolated, reducing habitat for fish that require wetlands for spawning and nursery habitat. The effects of water-level reductions in smaller lakes could be equally profound.

A more subtle and indirect effect of reduced stream inflows is potential lake warming. Inflowing streams transport dissolved organic carbon (DOC) from the terrestrial landscape into lakes. Aside from its importance as a nutrient, DOC absorbs the light energy passing through the lake's water column. A reduction in the delivery of DOC to lakes allows light to penetrate more deeply into the water column, thereby heating the lake to greater depths.

Thus, if streamflows are reduced (e.g., in summer), warmer (and less-oxygenated) water may extend to deeper levels in lakes. This reduces the cool-water habitat refuge of deep water that many fish require. Additionally, ultraviolet light can penetrate more effectively in surface waters where DOC is reduced. This harmful light can diminish algal productivity in surface waters, further clarifying the water, which allows light to penetrate even deeper and heat the lake.

Evidence of this effect of reduced inflows on lake ecosystem processes was documented empirically in western Ontario, Canada, between 1969 and 1989, a period of low precipitation, high evaporation, and subsequent low stream inflow to lakes. Lakes were not flushed as often with new water, and concentrations of chemicals in both lakes and streams increased. DOC levels dropped as well. As a result, dissolved oxygen levels dropped and some fish stocks were threatened.

Expansions and contractions of lake species

Warming of lakes will allow fish and other aquatic species that are currently restricted in their distribution by very cold water temperatures to expand their ranges. Under a doubling of CO_2, range expansions of warm-water species can be expected, maybe up to 500 kilometers north on average. However, the rate and extent of migration will depend on availability and access to dispersal corridors. Successful dispersal of fish depends on the degree to which lakes are interconnected via streams and the kinds of physical barriers that may exist in connecting streams.

In many glaciated regions, lakes are interconnected by streams and species are able to move across the landscape. In general, fishery yields decrease with increasing latitude, and because warming of northern lakes will increase available habitat for warm-water fish, these lakes may support a more diverse and productive fishery. However, the importance of interactions among new mixes of species is unknown, especially when large predatory species are included in the mix.

Freshwater Wetlands

Shallow freshwater wetlands are already stressed as ecological transition zones between aquatic and upland terrestrial ecosystems, making them particularly sensitive to changes in temperature and precipitation. Freshwater wetlands have been extensively altered by humans. More than half of the wetlands in the lower 48 states have been converted to other uses since the mid-1700s. Many of those that have not been converted exist in a degraded condition because of the cumulative effects of altered flow regimes, removal of timber, and deterioration of water quality due to human development and encroachment.

Because freshwater wetlands provide many critical goods and services (Table 1), they have become highly valued by large sectors of society; yet human activity still threatens these sensitive ecosystems. Climate change is expected to further stress wetlands, because increasing temperatures will affect species growth and reproduction, and altered water regimes will change soil and vegetation conditions.

Freshwater wetlands provide habitat for species that have adapted to widely varying soil moisture and chemical conditions. Purely wetland species are often habitat specialists that can survive nowhere else on the landscape. Further, wildlife populations are increasingly using wetlands as refuges as humans modify other types of ecosystems. For this and other reasons, wetlands are very important contributors to biodiversity and wildlife in most landscapes.

The driving force of wetland ecosystems is the hydrologic regime, because it regulates the "wetness" of the wetland. The patterns of water depth, and the duration, frequency, and seasonality of flooding together constitute a wetland's hydroperiod, which determines its vegetation composition, habitat for aquatic organisms, and other ecosystem characteristics. Viewing wetlands according to hydroperiod is perhaps the best way to consider potential effects of climate change. For example, in floodplain swamp forests, trees such as bald cypress are indicative of long periods of flooding. A shorter hydroperiod under drier conditions will allow species that are not as tolerant of prolonged soil saturation and oxygen deprivation to invade. By contrast, a longer hydroperiod normally results in tree death and eventual replacement by aquatic plants.

Carbon dioxide enrichment has the potential to alter species composition in some wetland types, independently of hydrologic or temperature changes. Some groups of plants are more responsive to higher CO_2 concentrations than others because of fundamental physiological differences. Marshes on the Chesapeake Bay, where most relevant studies have taken place, show quite striking increases in photosynthesis when CO_2 levels are raised. These studies suggest that, over the long term, the more responsive plant species may eventually crowd out the less responsive ones.

However, there are many caveats, not the least of which is the difficulty of scaling up from individual plant responses to whole ecosystem responses, where many variables other than just CO_2 are important. In addition, it is difficult to generalise about the effects of CO_2 on wetlands because of uncertainty about changes in other important factors such as water use efficiency, insect and fungal damage, and soil bacterial activity. Therefore, broad-scale generalisations of the effect of CO_2 enrichment on wetland communities are not currently feasible.

Responses to CO_2 enrichment are further masked by human-caused enrichment by nitrogen and phosphorus. It is common for aggressive plants (including non-native invasive species) to dominate when excess nutrients are available. Indeed, in low-fertility sites such as bogs and some lake shorelines, nutrient enrichment is known to favor weedy, aggressive species that can outcompete the native dominant species as well as rare and endangered species. In these settings, additional CO_2 enrichment may contribute to the loss of sensitive species and thus reduce regional plant biodiversity.

Consequences of altered water regime

The vulnerability of wetlands to a drying climate depends, in large part, on the sources of their water supply. The dominant sources of water to wetlands are precipitation, surface flows, and groundwater discharge. As a general rule, wetlands that are fed mainly by precipitation are the most likely to lose wetland characteristics in a drying climate. Groundwater-driven sites that have large volumes of available water stored in aquifers will have the greatest resistance to climate change. The responses of surface-flow wetlands will be somewhere between those types.

Wetlands dominated by precipitation are already at the "dry end" of the spectrum of wetland types. They tend to occur in flat landscapes with low soil drainage in humid climates, such as parts of the eastern United States and Alaska. If the climate were to dry, these wetlands would contract, resulting in the loss of locally unique species such as orchids and insectivorous plants characteristic of acidic peat bogs. Plant species less tolerant of flooding would dominate, and, over time, soils would dry to the point that they would no longer support the nutrient cycling processes unique to wetland soils. Peat-based wetlands would be especially hard-hit as the highly organic soils undergo oxidation and subsidence, thus altering drainage patterns, topography, and exposure to fire.

In a more humid climate, precipitationdominated wetlands could expand, assuming no barriers from competing land uses. The hydroperiods of many wetlands are driven by surface waters such as adjacent rivers and lakes. Climate change that alters hydrologic regimes of rivers and lakes will have ecological impacts on these wetlands. For example, a reduction in the frequency or magnitude of high flows that inundate the floodplain would tend to dry out floodplain wetlands, isolate them from the adjacent stream or river, and replace wetland plant species with more terrestrially adapted species.

Floodplain wetlands along rivers with a snowmelt hydrology may be particularly harmed, because peak flows from seasonal snowmelt may disappear. Analogues of this effect are seen below dams on impounded rivers, where reservoirs store flood flows and thus "shave" downstream flood peaks, transforming floodplain forests to species adapted to drier conditions. Lake fringe wetlands also tend to be driven by surface water, responding to both seasonal and interannual variations in lake water levels. Multiannual cycles cause the position of these types of wetlands to migrate back and forth across shallow shorelines.

Their susceptibility to climate change will depend largely on shoreline morphology. For example, deepening of water under a wetter climate would eliminate some wetland plant species, especially in areas where the shoreline is too steep to allow plants to become established. Lower water levels would require that plants become established further lakeward, but this could only happen if protective barrier beaches form along the shoreline. This is of particular importance for the Great Lakes, where lake levels might drop significantly under projected climate change.

Groundwater is the principal water supply for many wetlands. If climate becomes wetter, rising water tables will expand wetland areas and make existing wetlands even wetter. Under a drier climate, the opposite will occur. For example, declines in water tables in the prairie pothole region of the Dakotas will cause many seasonal wetlands that typically remain flooded well into the summer to dry out weeks earlier. This is the most important region in North America for breeding waterfowl. Using models that calibrate waterfowl populations and pond numbers to the Palmer Drought Severity Index and general circulation models (GCMs), Sorenson et al. predicted that breeding bird populations would be reduced to about 50 percent of present long-term averages in response to reduced precipitation and temperature increases of up to 4°C.

Drawdown of water for human uses provides additional insight into the effects of a drier climate on groundwater-dominated wetlands. In the northern Tampa Bay area of Florida, peat-based cypress domes have dried out because of drawdown for human consumptive use. This has caused the peat to oxidise and collapse, resulting in more destructive fires, invasion of weedy upland plants, and abnormally high tree-fall. Other factors come into play for wetlands in arid climates, most of which receive groundwater as their dominant source. For example, riparian cottonwood forests in the arid West have died as a result of groundwater pumping.

These human-induced changes, however, typically occur much more abruptly than those expected from a drier climate. Further, because many groundwater-supplied wetlands are supported by large aquifers, these wetland types may be affected less by climate change than wetlands supplied by precipitation and surface water. However, if conditions become drier, there is likely to be increased human demand for groundwater, which could negatively affect these wetland types.

Changes in temperature

Extensive regions throughout boreal and arctic zones of Alaska and Canada are covered by wetlands made up of organic-rich soils underlain with permafrost. The warming predicted for these regions will increase the depth at which the ground stays frozen (i.e., lower the permafrost table), resulting in drainage that will expose the peat to the atmosphere, thus increasing the rates of organic decomposition and the release of CO2 and methane. Given the probability of increased fire frequency and intensity that accompanies drier conditions, more peat will tend to burn under a warmer climate. Carbon dioxide and methane generated from these sources would stimulate a positive feedback to global warming.

Groundwater-dominated wetlands, or fens, are also an important habitat for many rare species of plants and animals. Because groundwater travels slowly through subsurface pathways, it is buffered from atmospheric temperature extremes. Thus, the thermal regimes of fens are much cooler in the summer and warmer in the winter than adjacent surface waters. Many fens support wetland plant and animal species that normally occur in abundance only much farther north, where summer water temperatures are cooler. These species were probably once common in the southern end of their current ranges when the climate was cooler, but they are now isolated. Thus, a change in future climate that eventually warms groundwaters will cause a loss of these species from the southern end of their ranges.

Changing plant and animal movements

The geographic distributions of plants and animals living in freshwater wetland ecosystems may shift in response to a changing climate. Because the types of wetlands defined above occur in all climates (except those dominated by precipitation), transformation to a warmer and drier (or wetter) climate in a particular region will produce conditions that already exist for wetlands in a similar climate at a different geographic location.

For example, silver maple/ash/elm forests of the upper Midwest could be replaced by cypress/tupelo swamps, currently limited to the Southeast and along the Mississippi Valley. Treeless tundra wetlands of the Alaska North Slope permafrost could be replaced by black spruce peatlands with discontinuous permafrost. In each case, colonisation by wetland plants and animals could be determined by the addition of species that migrate to their more favorable climatic conditions elsewhere and the removal of species that become locally extinct because of changing environmental conditions. Species that survive the changed conditions have traits that are suited to the new environment. To some extent these traits allow predictions of survival under future environmental conditions.

If all species had instantaneous dispersal powers in response to climate change, and enough was known about the tolerances of individual species, it might be possible to predict with great accuracy the responses of species to changing climate. However, such predictions are precluded by many confounding factors, including formidable barriers to migration (both natural and human-caused, such as dams and dikes), changes in frequency and seasonality of fires, and modification of water quality. Species vary greatly in their abilities to disperse, so the success of plants and animals in colonizing suitable habitats under a changing climate will be highly variable.

Many recently introduced, non-native species have expanded rapidly across the landscape, and analysis of their mode of colonisation may allow insight into which native species may be best able to disperse successfully under a changing climate. These invaders may be expected, however, to interfere with the gradual spread of indigenous wetland plants, because the invaders are able to rapidly expand their range and to monopolise available space and nutrients before indigenous plants can arrive.

Further, the proliferation of these exotic species is favored by human activities unrelated to climate change (e.g., altered hydrology, nutrient loading, and sedimentation). For some wetland species, geographic barriers are so formidable that they are unlikely to be overcome in the time frame projected for climate change. For example, the flora of alpine wetlands, restricted to the highest peaks in the continental United States, are particularly vulnerable, because an increase in temperature will eliminate species that require cold thermal regimes.

Coastal Ecosystems

Potential changes in temperature and hydrologic regimes projected to occur over the next 100 years will most likely lead to loss of coastal wetlands, deterioration of water quality, and disruptions in fisheries. Sea-level rise, sediment starvation, and changing species composition in coastal wetlands all threaten the sustainability and resilience of these systems. Coastal wetlands and estuaries are among the most productive ecosystems on earth. The high productivity of these systems is related to the inputs of freshwater and to the fluctuating water levels caused by the ocean tides.

Freshwater inputs to estuaries deliver not only nutrients that support high production, but also large quantities of suspended sediments, which are critical to the accretion of wetland soils that allow plants to avoid permanent inundation by the sea. In addition, the mixing of freshwater and seawater in estuaries creates water circulation patterns that tend to retain nutrients, which further enhances coastal ecosystem productivity. Coastal wetlands support dense populations of animals, many of them migratory and of great commercial value.

The movements of many migratory aquatic animals (e.g., crabs, fish) to deeper sea water are regulated by estuarine salinity, which changes as freshwater inflows mix with tidal salt water. Temperature conditions are also an important environmental cue for migration. Further, water temperature regulates many ecological processes and the distribution of many coastal wetland species. Coastal wetland ecosystems are also among the most altered and threatened natural systems.

Important impacts include increased nutrient loading leading to eutrophication, direct loss through habitat destruction, changes in hydrology, introduction of toxic materials, and changes in species composition due to over-harvest and introduction of new species. Most coastal wetland loss has been due to draining and filling. In some coastal states, such as California, almost all coastal wetlands have been lost. In the largest coastal ecosystem in the United States, the Mississippi Delta, restriction of river sediment input is largely responsible for massive wetland loss.

Coastal eutrophication is a growing problem because of increasing inputs of nutrients from agriculture, industry, and human populations. The combined effects of climate change, wetland loss, and eutrophication will be worse for wetlands than the impacts of these individual stresses acting alone. Coastal wetland plants thrive in a relatively harsh intertidal environment characterised by alternate flooding and draining of salt marshes with associated waterlogging of soils, depletion of oxygen, and production of natural toxins that inhibit plant growth.

To cope with these harsh conditions, plants have a number of adaptations, including the production of aerial roots and submerged tissues that allow them to capture oxygen needed by the roots. But these adaptations are suitable only as long as the average water level remains relatively constant. Accordingly, coastal wetlands exist within a fixed elevational range, where the frequency and duration of inundation by seawater are relatively constant. Because plants become progressively more stressed and ultimately die if they are inundated for too long, an increase in water levels due to sea-level rise can severely stress the integrity of coastal wetland ecosystems.

Sea-level rise over the last several decades has reportedly led to salinity intrusion and wetland loss in a number of coastal areas around the world and in the United States, including Long Island, the mid-Atlantic region, Chesapeake Bay, and the Mississippi Delta. Since sea-level rise over the last century is two to nine times lower than the projected 20-90 centimeters of sea-level rise expected over the next 100 years, there is great concern for the potential loss of coastal wetlands in the United States and globally.

The projected rise in sea levels under global climate change will certainly place these productive and important ecosystems under additional stress, with the very likely result of extensive dieback of plants residing in the current intertidal zone. For example, a 0.5 meter rise in sea level would inundate about 12,000 square kilometers (about 4,600 square miles) of coastal wetlands, most of which would likely be lost.

Coastal Wetland Survival

For coastal wetlands to persist in the face of projected sea-level rise, the soil surface must accrete vertically at a rate two to nine times that observed over the last century. During periods of sealevel rise, coastal wetlands can persist only when they accrete soil vertically at a rate at least equal to water-level rise. A number of studies have shown that coastal marshes are indeed able to accrete at a rate equal to the historical rate of sea-level rise and persist for hundreds to thousands of years. Given that sea level is projected to increase up to two to nine millimeters per year over the next 100 years, soil accretion will have to occur at a much higher rate than that observed over the last century.

In the Mississippi Delta, for example, where there is sufficient sediment input from the river, accretion rates greater than 10 millimeters per year have been measured. Such accretion rates are possible in other systems as well, if sufficient sediment is available. The rate at which accretion occurs is a function of the combination of the inputs of both inorganic and organic material to the soil. Organic material is mostly derived from the growth of plant roots, whereas inorganic material is mostly supplied in the form of sediments that come from either the sea or freshwater sources.

Riverine sediments are generally more important because their input is more frequent and they contain nutrients that enhance organic soil formation. Many rivers emptying into coastal estuaries now carry only a fraction of the inorganic sediment that they did historically. For example, sediment discharge to the Mississippi Delta has decreased by over 50 percent since 1860, largely due to the building of Missouri River dams, which capture and store the suspended sediment in the river water, resulting in a significant loss of coastal wetlands.

Although wetlands are able to migrate inland as sea levels rise, in many cases barriers of human development exist that preclude effective migration. Also, rapid increases in elevation often characterise the uplands bordering coastal wetlands, which means that upland migration could support only a small fraction of the wetlands expected to be lost to rising sea levels.

Changing freshwater input to coastal systems

Changes in freshwater runoff to coastal wetlands will alter the availability of sediment needed to accrete soils as sea levels rise. For much of the U.S. coast, most GCMs agree that winter and spring rainfall will increase, although there is disagreement among models as to whether precipitation will increase or decrease in summer and fall. Thus, there is the potential that freshwater discharge to coastal systems will increase during some seasons but decrease during others.

Increased freshwater input can have both beneficial and detrimental impacts on coastal systems. The benefits associated with vertical soil accretion have already been highlighted. An additional benefit could be an increase in fisheries production in coastal systems. This could result because the nutrients in freshwater flowing into estuaries stimulate primary production, which in turn increases the energy available for organisms on which fish forage. In the Mississippi Delta, these diversions add sediments that increase accretion, lower salinity to combat saltwater intrusion, and benefit fisheries and wildlife.

There is concern, however, that diversions will lead to algal blooms because of the added nutrients and that they will add pollutants, which may severely affect organisms in these areas. Thus, diversions will have to be studied and managed carefully to avoid problems. On the other hand, decreased freshwater inputs are likely to lead to less accretion, lowered productivity, and saltwater intrusion. One potential negative impact associated with an increase in freshwater runoff to coastal ecosystems is an excessive increase in nutrients.

There is already considerable evidence that agricultural runoff and wastewater from human activity in tributary watersheds are degrading many coastal ecosystems. Nuisance algal blooms and low oxygen in bottom waters kill fish and shellfish, as has been well documented in the Chesapeake Bay and off the coast of Louisiana in the so-called Gulf Dead Zone. The increased runoff will also lead to problems with toxic pollutants (e.g., heavy metals, organic chemicals) if there are high levels of these chemicals in the runoff.

Several suggested changes in environmental management have been made to reduce high nutrient loading to streams and coastal waters. For the Mississippi River basin, for example, these recommendations include changes in farming practices, use of buffer strips along steams, protection of wetlands to improve water quality, and reduction of nitrate in river water by diversions in the Mississippi Delta.

Consequences of higher temperatures on coastal vegetation and fisheries

With a warming of waters in coastal wetlands, substantial shifts in species composition are expected. One important likely result of increasing temperatures along the southeast U.S. coast will be a northward migration of mangroves, which will replace salt marshes. Mangroves are tropical coastal forests that are frost-intolerant. Chen and Twilley developed a model of mangrove response to the frequency of freezes. They found that when freezes occurred more often than once every eight years, mangrove forests could not survive. At a frequency of 12 years, mangroves replaced salt marsh. Along the Louisiana coast, freezes historically occurred about every four years. By the spring of 2001, however, a killing freeze had not occurred for 13 years, and small mangroves are found over a large area. If this warming trend continues, mangroves will spread even further, over much of the northern Gulf and part of the South Atlantic coast.

Because mangroves have many of the same ecological functions as salt marshes (high productivity, habitat for wildlife and fish, sites of nutrient uptake, etc.), a switch in U.S. coastal wetlands from salt marshes to mangroves might not change ecosystem function. However, if the climate becomes more variable, with frost-free periods interspersed with occasional hard freezes, it could be more difficult for either marshes or mangroves to survive, resulting in a loss of wetland habitat. Further, mangrove forests are more prone to damage by hurricanes than are marshes, so the stability of coastal ecosystems might diminish as mangroves spread.

Global climate change and sea-level rise can influence coastal fisheries in a number of ways. Approximately 70 percent of the U.S. fisheries' catch is derived from estuarine-dependent species, and their young are therefore dependent on suitable habitat. Rising sea levels that lead to destruction of coastal wetlands can have direct negative consequences for coastal fisheries. More indirect effects could result from wetland loss that leads to shoreline erosion, which, by adding fine sediments to the water column, would reduce water clarity and thus interfere with feeding ability. Many shellfish species also use coastal wetlands as an important habitat; therefore, they are vulnerable to wetland loss caused by sealevel rise.

Fish and shellfish species that use coastal wetlands as nursery habitat are also sensitive to temperature conditions. As climate warms, many species will be forced to shift their geographic ranges northward toward suitable thermal rearing environments for their young. The continued production of coastal fisheries will then be partially dependent on these species finding coastal environments containing functional nursery habitats at the higher latitude. Juvenile shellfish and fish rely on environmental cues for migration out to the sea, and these cues might be masked by changes brought on by the direct and indirect consequences of climate change.

For example, large changes in freshwater inputs could also alter salinity gradients and estuarine circulation patterns that are important cues for migratory fish and shellfish. Changes in coastal ocean circulation patterns caused by changes in freshwater inputs and temperature may lead to changes in the regional patterns and spatial distribution of production as well as variability in naturally fluctuating stocks such as herring and sardines, thus affecting national and local economies. How offshore waters will respond to climate change is poorly understood.

Changing storm intensity, frequency, and track

The frequency, track, duration, and intensity of storms and hurricanes might also change with global climate change. There is some empirical evidence that the frequency of Atlantic hurricanes increases with increasing sea-surface temperatures, although the uncertainty in such predictions is high. In addition, with warmer sea-surface temperatures, hurricanes might reach higher latitudes more often. Storms may also be accompanied by greater rainfall intensity. Increases in hurricane intensity predicted by the models, however, fall within the range of natural interannual variability and of uncertainties of current studies.

In general, long-term changes in the frequency, intensity, timing, and distribution of strong storms would most likely alter the species composition of coastal marshes, as well as the rates of important ecosystem processes such as nutrient cycling and primary and secondary productivity. For coastal systems of the southeastern United States, there could be both positive and negative effects. For example, hurricanes greatly increase the rate of soil accretion in marshes, thereby helping to offset accelerated sea-level rise. Runoff generated by hurricanes introduce freshwater and nutrients that can enhance coastal wetland productivity. In the arid areas of south Texas, freshwater input can also have a stimulatory impact by reducing salinity and its attendant stresses.

On the negative side, hurricanes can reduce the structural complexity of coastal forested wetlands such as mangroves and tidal freshwater forested wetlands. The locations with the highest probability of hurricane landfall are south Florida and the Mississippi Delta. The coastal ecosystems of south Florida are dominated by mangroves and, if warming continues, the Mississippi Delta will become increasingly dominated by mangroves. This trend, combined with increased hurricane frequency, would reduce the structure of these forests and destroy some of them. Freshwater forested wetlands (swamps) of the Mississippi Delta are slowly degrading and disappearing because increased flooding from rising water levels has largely eliminated the establishment and growth of young trees.

Hurricanes can also cause tree loss; for example, Rybczyk et al. reported that nearly 10 percent of trees in a swamp forest in Louisiana were blown down during the passage of Hurricane Andrew in 1992. An increase in hurricanes would amplify the kind of damage resulting from the interaction between rising water levels and hurricanes and would hasten the loss of forests in the Mississippi Delta and elsewhere. High runoff from hurricanes can also lead to excessive nutrient loading and eutrophication problems. For example, record runoff from Hurricane Floyd into the Pamlico Sound estuary in North Carolina led to water quality problems.

Ecosystem Adaptation

For many reasons, the ability of aquatic ecosystems to adapt to climate change is limited. Expected rates of climate change are probably too great to allow adaptation through natural genetic selection. Many types of habitat will be diminished or possibly lost entirely (e.g., alpine wetlands). Animals and plants will need to disperse northward or to higher elevations, but aquatic species differ greatly in their dispersal abilities, so not all species will be able to move to hospitable habitat.

Further, most high-quality aquatic habitats are now spatially isolated (due to human activities), making successful dispersal even more difficult. The extinction of many local species can be expected across all aquatic ecosystem types. Many surprises are also expected as species migrate across the landscape and come into contact with new species for the first time, particularly exotic species that humans have introduced outside their historical ranges.

In general, there is strong scientific consensus that aquatic ecosystems are vulnerable to projected climate change. However, the amount of ecosystem alteration directly attributable to climate change will be difficult to ascertain, because other powerful agents of global change (e.g., human activities, exotic species) will continue to independently and interactively degrade aquatic ecosystems. One critical uncertainty in projecting future aquatic ecosystem response to a changing climate is how humans will interact with these ecosystems as climate changes.

Human activities have severely modified many aquatic ecosystems with actions such as diversion, groundwater pumping, and the building of dikes, levees, and reservoirs, all of which have modified natural processes and increased the vulnerability of aquatic ecosystems to additional stress associated with climate change. Cumulatively, these alterations fragment the aquatic landscape and make dispersal between ecosystems more difficult.

A rapidly changing climate introduces uncertainty in water resource management and therefore threatens aquatic ecosystems if they are not adequately considered in human adaptation to climate change . To date, the needs of aquatic ecosystems have not generally been adequately represented in the formulation of environmental policy, especially the need for water quantity and timing that can sustain aquatic biodiversity and productivity.

Thus, a challenge for human society will be to adapt in ways that minimise damage to these important ecosystems. Human activities should maximise the potential for adaptation of these ecosystems by minimising such present environmental stresses as pollution, habitat destruction, fragmentation, and exotic species introduction. These and other actions could enhance aquatic and wetland ecosystems, regardless of whether climate changes with the projected magnitude or not. Climate change offers additional incentives to implement more science-based management of aquatic resources; this could be considered "climate change insurance."

References

Clark, W.G. & Hare, S.R. 2002. Effects of climate and stock size on recruitment and growth of Pacific halibut. *N. Amer. J. Fisher. Manag.,* 22: 852.

Clemmensen, C., Potrykus, A. & Schmidt, J. 2007. *Climate change and European fisheries.* European Parliament, policy department, structural and cohesion policies, fisheries. Brussels. Study.

Croxall, J.P., Trathan, P.N. & Murphy, E.J. 2002. Environmental change and Antarctic seabird populations. *Science,* 297: 1510.

Edwards, M. & Richardson, A.J. 2004. Impact of climate change on marine pelagic phenology and trophic mismatch. *Nature*, 430: 881.

11

Agricultural Adaptation to Climate Change

Climate change the ultimate global warming impact due to green house effect is changing most of our traditional agricultural practices as the seasonal cycle and rainfall pattern have changed, droughts have become more frequent, violent stresses of natural disasters like cyclones, earth quakes, prolonged floods, salt water intrusions are increasing day by day. The average temperature has increased in the summer while winter season has shortened.

As a consequence 17 percent of total cultivable land would be affected. Three crore of people would lose their farms and homesteads. Threatening the richest and most productive region of the country, sea level rise could have dramatic consequences for its economy. Increased flooding from glacial melt, more intense monsoons, or more intense cyclones could also adversely affect agriculture in near term, periodically inundating much agricultural land. Sea level rise under climate change would also result in saline intrusion into the river system, which would enhance the backwater effect. Several hundred thousand tons of grain production could be lost as a result of increased salinization from sea level rise.

An early part of adapting agriculture to climate change involves helping communities to understand why adaptation is a needed part of today's vision of the future and therefore of their management strategies. Successful adaptation to climate change will require flexible, risk-based approaches that deal with future uncertainty and provide strategies that are robust enough to cope with a range of possible local climate outcomes and variations.

Many climate adaptation options for agriculture are similar to existing 'best practice' and good natural resource management, and do not require farmers to make radical changes to their operations and industries in the near term. These options can, and should be, prioritised as part of a 'no regrets' or win–win strategy for agriculture because they will provide immediate and ongoing benefits as well as preparing the sector for climate change.

Higher temperatures will mean crops that poor people depend upon will no longer grow in some areas. Some regions will experience less rainfall while others more. The frequency of droughts and floods will increase, water flows will become more unpredictable, and coastal food-producing areas will be inundated. In regions receiving more rainfall, increases in temperature will increase water demand from crops because of higher evapo-transpiration. These will all have major impacts on food security, food prices and malnutrition in the developing world.The poorest of the poor will be the hardest hit, especially as agriculture is the main source of living for the majority of these people. Loss of agricultural land and competition for resources, especially water, has the potential to cause large scale migration and conflict.

Vulnerability to food security has (difficult phase in) grown global. Global and local food security vulnerability patterns will be modified by climate change. Small-scale rainfed farming systems, pastoralist systems, inland andcoastal fishing and aquaculture communities and forest-based systems are particularly vulnerable to climatechange. It is imperative to improve preparedness to futureand uncertain impacts through preventive and planned adaptation and innovation, technical adaptation measures range from change in production systems like adjustingplanting or fishing dates, rotations, multiple cropping/species diversification, crop-livestock pisciculture systems,agroforestry investing in soil, water and biodiversity con-servation and development by building soil biomass, restoring degraded lands, rehabilitating rangelands, harvesting and recycling water, planting trees, developing adaptedcultivars and breeds, protecting aquatic ecosystems tomaintain long-term productivity. Adaptation measures alsotake account of establishing disaster risk management plans and risk transfer mechanisms, such as crop insuranceand diversified livelihood systems.Mitigation options include carbon sequestration in agriculture and forestry. Mitigation of climate change is a global responsibility. Agriculture, forestry, fisheries/

aquacultureprovide in principle, a significant potential for greenghousegases mitigation. The IPCC estimates that the global tech-nical mitigation potential for agriculture will be between 5,500 and 6,000 Mt CO_2-equivalent per year by 2030,89% of which are assumed to be from carbon sequestra-tion in soils. The potential benefits of carbon sequestration are:

i) Mitigation is done when CO_2 is removed from theatmosphere;

ii) Adaptation is achieved when higher organicmatter levels in soil increase agroecosystem resilience and

iii) Income generation and livelihood is sustained whenimproved soil fertility leads to better yields.

Major invcstmcnts in adapting agriculture to climate change are therefore going to be needed to build resilience and protect the poor. This will need to be alongside measures to mitigate GHG emissions, and isolate carbon in agricultural systems. Proposed mechanisms for addressing GHG emissions from agriculture include mitigation financing under NAMAs (Nationally Appropriate Mitigation Actions) and reforms to the Clean Development Mechanism (CDM). Investments in adaptation now need to avoid "locking" countries into a high emissions route. An understanding is also needed of optimal adaptation and mitigation points. This is to know when more radical measures smay be needed such as migration and/or leaving the agriculture sector when it is no longer a viable livelihood route.

High-input Agriculture

The world needs more grain for food. This has to be achieved even with diminishing arable lands, water scarcity, and rising costs of cultivation. In addition, climate change is adding further complexity. The world population of 6.07 billion in 2000 is projected to grow to 8.13 billion in 2030 and to 8.92 billion in 2050. Average per capita food consumption in developing countries, which rose from 2,110 kcal person^{-1} day^{-1} 30 years ago to 2,650 kcal person^{-1} day^{-1} at present, may rise further to 2,960 kcal person^{-1} day^{-1} in the next 30 years and to 3,070 kcal person^{-1} day^{-1} by 2050. The Food and Agriculture Organization (FAO) and the World Food Program (WFP) reported in their 2009 publication, *The State of Food Insecurity in the World*, that, for the frst time since 1970, more than a billion people worldwide (around one-sixth of all humanity) do not have enough food and are hungry and undernourished.

World agriculture has grown at 2.1% to 2.3% per annum in the last four decades. The future may see a drastic decline in the growth of aggregate world production to 1.5% per annum in the next three decades and to 0.9% per annum in the subsequent years to 2050. The cereals sector (sum of wheat, milled rice, and coarse grains) has been in a downward trend for some time now, with growth rate falling from 3.7% per annum in the 1960s to 2.5%, 1.4%, and 1.1% in the subsequent three decades to 2001. An increase in world production by another 1.1 billion t annually will be required by 2050 over the 1.9 billion t of 1999-2001 (or 1 billion t over the 2 billion of 2005). Achieving it should not be taken for granted as land and water resources are now more stretched than in the past and the potential for continued growth of yield is more limited. A declining trend in crop yield and an increasing water shortage are apparent in many nations.

Since the beginning of the 1960s, the large increase in the demand for food has been met largely by improved agricultural productivity, which required more inputs—fertilizers, pesticides, and improved seeds—and more water for irrigation. Agriculture demanded more and imposing more inputs, putting pressure on water resources and ecosystems and imposing a heavier fnancial burden on many countries. It is estimated that 70–80% of future increases in crop production in developing countries will have to come from intensifcation, that is, higher yields.

India, having an estimated population of 1.4 billion by 2025, will require 300 million t of grain compared with the approximately 200 million t at present. Little extra land is available and the increase in production will have to come from higher yields, for which there is ample scope.

The cost of cultivating food crops has consistently increased owing to the escalating costs of seeds, fertilizers, and labor. With increasing scarcity of labor because of urbanization, sustaining the interest of farmers in crop production itself has become a challenge.

Zeigler et al listed seven challenges to be addressed in intensive irrigated rice systems. Two of these are (1) exploiting all options for raising the yield potential of rice and (2) closing "yield gaps," increasing yield stability, and improving net returns through improved germplasm with multiple resistance to abiotic and biotic stresses and improved crop management. The projected breakthrough by modifying rice photosynthesis through a C_4 pathway will take millions of dollars and at least 10 to 15 years of dedicated work by a global scientifc team (Zeigler et al 2008). So far, the much-publicized strategies surrounding genetically modifed (GM) rice have not had much success as seen by the limited spread of the "new plant type" (NPT).

High-input agriculture has clearly reached its limit. Even the more complex solutions in the pipeline require large fnancial resources, and, if such resources are made available, they require more than 15 years to reach farmers. There is clearly an urgent need to fnd ways to grow more food with less water and fewer inputs. Farm-based approaches, which used to be at the center of agricultural practices for centuries to improve productivity, need to be explored once again. Farm-based approaches are relatively easy and the results are visible in a short period of time. Some new approaches clearly show promising results in farmers' felds. These methods require further improvement and refnement not just through investments in the feld but also through research.

Increasing Productivity

Current productivity of most smallholder farmers is still far below what is possible and routinely achieved in countries where investment has been appropriate. A marginal increase in smallholder farm productivity worldwide will substantially improve food security. These smallholders, however, may not be able to access the complicated and expensive solutions that are currently in the market or in the pipeline. There are indications in India that these expensive solutions actually lead farmers to commit suicide. This is because of the debts that they incur from purchasing these inputs. Clearly, for these smallholders, the "more with more" approach is the end of the road and even the end of their lives. Unsustainable land and water management practices, including deforestation, have also contributed to losses in soil fertility and productivity and disruptions in food production and economic development, especially in the most fragile and marginal environments, where smallholder farmers are the major custodians of natural resources. Unleashing the full potential of smallholder farming is key to the global food security agenda.

Globally, there is a large reserve of unused potential farmland. However, only a fraction of this land is realistically available for agricultural expansion as much of it is needed for other purposes, such as forest cover and infrastructure development. In West Asia and North Africa, at least 87%

of the suitable land has already been farmed; in South Asia, the corresponding fgure is 94%. In many areas, land degradation threatens the productivity of existing farmland and pasture. Even if there is land for farming globally, that is not going to solve the problem. Land in countries where people reside has reached its limit. There is no more new farmland in countries where food security is an issue. And, even if it exists, it is still expensive to farm.

From 2001 to 2006, world grain production grew by just 1% per annum, whereas world grain production per capita has continued to fall by –1.2% per annum compared with –0.3% per annum in the 1990s. When the Green Revolution started in China, 1 kg input of N could contribute to 15–20 kg output of additional rice. Today, that increment is only about 5 kg, and it continues to decline.

Under climate change, crops in many regions are prone to environmental stresses that have not been observed before. Many annual crops such as wheat, soybean, and rice have a threshold temperature above which seeds do not form properly. A brief episode of hot temperature (>32–36 °C) can devastate crop yields.

Grain demand in India is estimated to be about 300 million t per annum by 2020, necessitating an increase of about 91 million t from the estimated 209 million t production for 2005-06. Since there is no probability of any further increase in the area under cultivation over the present 142 million ha, much of the desired increase in grain production has to be attained by enhancing productivity per unit area. The productivity of milled rice has to be increased from the present 2,077 kg ha^{-1} to 2,895 kg ha^{-1} by 2020, with an average increase of about 5% per annum. The productivity of wheat has to be increased from the present 2,713 kg ha^{-1} to 3,918 kg ha^{-1}, with an average increase of about 7.5% per annum, whereas the productivity of pulses has to be increased from the present 637 kg ha^{-1} to 1,282 kg ha^{-1}, with an average increase of about 5.3% per annum. On the contrary, the productivity of most crops, except that of wheat, has shown a negative growth rate of 0.72% to 1.84% per annum in 2000-01 and 2002-03. This poses not only a matter of great concern but also a formidable challenge.

Grain production in India, following the Green Revolution in 1969-70, yielded 99.5 million t and it nearly doubled by the end of the last century. The highest average annual increase of 6.1% in grain production was recorded during the 1980s: from 110 million t in 1979-80 to 171 million t in 1989-90; but the annual increase in grain production during the 1990s dropped to 1.5%. The fact that fertilizer was the key input in augmenting grain production, next to the availability of seeds of high-yielding crop varieties, is evident from the increase in fertilizer (N + P_2O_5 + K_2O) consumption: from 1.98 million t in 1969-70 to 18.07 million t by 1999-2000. Nevertheless, the average annual increase in fertilizer consumption witnessed a declining trend in these three decades—16.5% in the 1970s, 12.04% in the 1980s, and only 5.6% in the 1990s. A simple regression analysis between grain production and fertilizer consumption from 1960-61 to 1999-2000 showed that the partial factor productivity of fertilizers has been continuously declining. This is supported by the fact that farmers in the rice-wheat cropping system belt (especially Punjab, Haryana,and Western Uttar Pradesh) are forced to apply more and more fertilizer to obtain the same crop yield of the preceding years.

There was no yield gain in wheat in the last 10 years and plant breeding is strained. An examination of the yield performance of advanced materials in the All-India Coordinated Wheat Improvement Project (AICWIP) multilocation yield trials in timely sown conditions showed that, during the last decade, no entry yielded statistically higher than PBW343.

Modern input-intensive rice farming typically involves monocrops, irrigation, and reliance on new seed varieties, synthetic fertilizers, and pesticides. This approach depends on large amounts of water and nonrenewable fossil fuels and has become associated with adverse affects on people's health and the environment over time. Since the dramatic increase in rice production that was achieved in the 1970s and 1980s, yield growth rates in many countries have slowed down. There has been no signifcant increase in the productivity of major food crops such as rice, wheat, maize, millet, sorghum, and pigeon pea in the past two decades.

Increased oil price at nearly $100 per barrel has led to a tremendous stress in government fertilizer subsidies costing, for instance, the Indian government in excess of $1 billion a year. To add to the increased input cost of agriculture, diminishing returns have been noted.

In the next two decades, water scarcity will increase dramatically in many parts of the world. This will have signifcant social and economic repercussions. Global grain harvests will be threatened, more countries will rely on food imports, and the livelihoods of many people will be threatened.

Water scarcity will have a signifcant impact on agriculture. Shortage of water for agricultural production has already become a major problem in some countries. In the world's "rice bowls"—particularly China and India—the scarcity of water is acute, with competing demands on freshwater sources triggering conficts. The declining freshwater availability will have its impact on agriculture in India.

For the last fve decades, focus on improving the productivity of crops is based on one thing: more inputs to get more. Farmers were told to use more water, more fertilizers, more (expensive) seeds, and more pesticides in order to produce more. But, giving more to get more has certainly reached its limit. The result is a crisis in the farm sector in many countries. Now, there is a need to think differently. The approach will be to fnd ways to produce more with less. In the context of climate change and uncertainty in resources such as water, there is a need for a fundamental shift from high-input, high-ecological-footprint agriculture to a more sustainable approach. From an era of high-input agriculture, it is necessary to plunge into low-input agriculture with less use of water, labor, and chemicals. Ensuring food security with this is a big challenge. Historical evidence shows that simple changes in agronomic practices could bring about major impacts on crop yield. Jethro Tull, a famous English farmer (1680-1740), jumped to the conclusion that tillage alone would serve instead of manure. Tull's principle was carried out to better end by Rev. Smith of Lois-Weedon of Northamptonshire. Operating upon a clay soil, Smith produced large wheat crops. His average for many years was 34 bushels instead of 16 bushels, which was the average yield of the locality. He used no manure but simply parceled out his felds in 5-ft-wide strips and grew the crop in drills on alternate strips in successive years. The vacant strips were plowed deeply and frequently, so that through disintegration of the soil, absorption of CO_2, and combined N from the air, plant food enough for the next year's crop was secured.

A century ago, an innovative farmer in Tamil Nadu, India, had the idea of modifying existing agronomic practices in rice cultivation with single seedlings, wider spacing, and some intercultivation operation and reported a yield of 6,004 kg ha^{-1}. This *gaja* method employed interrow spacing of 45 cm and within-row spacing of 30 cm between single plants, resulting in a plant population of only 7–8 plants m^{-2}. Further research into the history of rice cultivation in Tamil Nadu has revealed that, in 1911, several farmers published articles in Tamil language on single-seedling planting. Scanned copies of these articles in Tamil and their English versions have been published separately. It was also found that this single-seedling planting was popularized by the then British government in the Madras presidency.

Vaidyalingam Pillai's reported yield of 6,004 kg grain ha^{-1} with the *gaja* method in Thanjavur District was 2.7 times more than that obtained from the same feld the previous year using bunch planting. Yanagisawa has estimated that the average rice yield in Thanjavur District during 1911 was 1,693 kg paddy ha^{-1}, while the average yield in this district for 1911-15 was 1,492 kg ha^{-1}. It is fascinating to see that such high yields were being obtained by farmers with their own innovations a century ago, when no chemical fertilizers were applied. By 1914, single-seedling planting was being adopted on 40,468 ha.

A major aspect of maximum land use essentially concerns the cultivator himself. It involves improved methods of husbandry, such as manuring, use of better seed, improvement of cultural practices, and control of pests and diseases.

An exciting approach has recently been developed—the System of Rice Intensifcation (SRI) not only reduces the use of irrigation water but also increases yields signifcantly and enhances the livelihood of rice farmers. SRI is perhaps the best example of options available to farmers and nations to promote community-led agricultural growth while managing soil and water resources more sustainably and even enhancing their future productive capacity. As SRI modifes how farmers manage their plants but not the plants themselves, it is compatible with genetic improvement strategies while mitigating the drawbacks associated with monocultures, agrochemicals, and climate change. This makes it a win-win proposition for rural households, countries, and the planet.

SRI was introduced in India in 2000. Today, SRI is known in all rice-growing states in India. It is estimated that as many as 600,000 farmers are growing their rice with all or most of the recommended SRI crop management practices on about 1 million ha distributed across more than 300 of the country's 564 rice-growing districts. This is probably the most rapid uptake of new agricultural practices seen in the country, making SRI a national phenomenon with very limited resources devoted to extension. Both on-farm and on-station evaluations across many states and diverse growing environments have shown clearly that SRI has the potential to improve yield while reducing water use, production costs, and chemical inputs. Available data from SRI experiments across India show an increase in grain yield of up to 68%.

SRI methods are seen to have the following impacts compared with their conventional counterparts.

- Depending on current yield, output per hectare is increased usually by 50% or more, with increases of at least 20%, and sometimes 200% or more.
- Since SRI felds are not kept continuously fooded, water requirements are reduced, generally by 25–50%.
- The system does not require purchase of new varieties of seed, chemical fertilizer, or agrochemical inputs, although commercial inputs can be used with SRI methods.
- The minimal capital costs make SRI methods more accessible to poor farmers, who do not need to borrow money or go into debt.
- Costs of production usually decline by 10–20%. This percentage varies according to the input intensity of farmers' current production.
- With increased output and reduced costs, farmers' net income increases by more than their augmentation of yield.

In a systematically conducted experiment at Hyderabad in the rabi season of 2009-10, SRI crops under organic and inorganic nutrient management resulted in 8.1 and 8.2 t ha^{-1} grain yield with a 12.6% and 13.7% yield increase over the control, respectively. Careful measurement of water use showed that water productivity is higher under SRI and water savings reached 37.5% and 34.2% under SRI-organic and SRI, respectively.

Experimental evidence is forthcoming on why the rice crop under SRI results in higher yields than under conventional or recommended cultivation practices. Thakur et al showed that alterations in management practices can induce multiple, signifcant, and positive changes in phenotype from a given rice genotype. The increase in yield with SRI when compared with that obtained using recommended management practices reached 42% and it was associated with various phenotypic alterations such as longer panicles, more grains panicle, higher percent of grain flling, increased productivity per plant, deeper and better distributed root systems, higher xylem exudation rates, more open plant architecture with more erect and larger leaves, more light interception, higher leaf chlorophyll content at ripening stage, delayed senescence and greater fuorescence effciency, higher photosynthesis rate, and lower transpiration. The combination of transplanting single seedlings per hill and the following intermittent irrigation during vegetative growth stage improved root-length density and root activity rate as well as shoot growth and delayed senescence of plants, leading to higher grain yield. Zhao et al reported 26.4% higher yield and signifcantly higher microbial biomass and microbial biomass nitrogen under SRI when compared with traditional fooding. Yield increased because of the increase in chlorophyll (delayed leaf senescence) and biomass accumulation at later stages.

Farmers see that the main advantages of adopting SRI are considerable savings in seed, water savings of up to 50%, improved soil health, and yield increases of 20–30%. Additional benefts include shorter time to maturity, higher outturn of polished rice when SRI paddy is milled, and resistance to drought and storm damage. The major constraints experienced are lack of trained labor, diffculties in planting young seedlings, water management in low-lying areas, and greater requirements for weeding.

With climate change, increasing variability of rainfall, and the growing competition for water and land, SRI offers a new opportunity for increasing the production value per drop of water and for reducing agricultural water demand, which, in many parts of the world, accounts for the largest share.

In India, the signifcance of SRI is not limited to rice alone. Its core practices are applicable to other crops such as sugarcane. WWF with ICRISAT recently published a detailed manual, *Sustainable Sugarcane Initiative (SSI): improving sugarcane cultivation in India*. Demonstration sites are being put up in fve states. The initial results are excellent. Like SRI, SSI could have signifcant implications for the way that sugarcane is cultivated in the world. Following the principles of SRI, the System of Wheat Intensifcation (SWI) is gaining popularity in Himachal Pradesh and Uttarakhand states of India. SRI principles of lower seed rate, limited water use, and intercultivation are being applied to crops such as fnger millet (*ragi*), mustard, and pigeon pea in various states of India.

The world needs more grain for food. Current methods to improve yield have reached their limits. The "more for more" approach faces a major crisis so there is a need for a major shift in agricultural research. In addition to this, climate change is going to add to the uncertainty. Water shortage is already a major problem that limits rice production and climate change is going to add to the uncertainty of water source in many parts of the rice-growing world. Farm-based methods, which used to be the major focus before high-input agriculture came into the picture, need attention to improve productivity. Already, some practices such as SRI, SSI, and SWI are showing positive results in farmers' felds. These methods have attracted great attention from governments, aid agencies, and civil society organizations. These farm-based approaches highlight the fact that, without investing heavily in new irrigation projects and in modifying existing irrigation systems, water could be effectively used with higher productivity.

Recognizing smart water management and planting practices, farmers in Tamil Nadu have increased rice yields by 30–80% and reduced water use by 30%. World Bank President Robert Zoellick has emphasized that SRI addresses not only food security but also water scarcity, which climate change further aggravates. However, these methods still have not been adopted by mainstream research organizations. These methods are very promising, but they require further research and refnement to realize their full potential. The challenge of food security, worsened by the complexity of climate change, could be effectively addressed through farm-based methods, but they need large investments and require immediate attention.

Coping with Adverse Conditions

Unfavorable rice environments are defned as environments where rice production is frequently constrained by abiotic stresses such as drought, submergence, and adverse soil conditions. Most of these conditions are found where rice production is dependent on rain only (rainfed rice), but they also occur in irrigated or partially irrigated systems. The most important rice-based rainfed system in Asia, with respect to area and number of dependent households, is the rainfed lowland ecosystem, which covers about 46 million ha, almost 30% of the total rice area worldwide. Other rainfed systems with often multiple abiotic limitations are upland rice (about 8.9 million ha in Asia) and deepwater rice (about 3.7 million ha in Asia). Considerable areas of unfavorable rice environments in irrigated systems are constrained by submergence, soil salinity, acid sulfate soils, and peat soils. Many of these rice areas are located in deltaic and coastal regions. The most important abiotic stresses affecting rice production are drought (about 23 million ha regularly affected), submergence (about 20 million ha regularly affected), and salinity (about 15 million ha affected).

Thus, weather- and soil-related abiotic stresses are already widespread in many rice-based cropping systems. Climate change, which is predicted to cause higher temperatures, more extreme rainfall events, and a considerable sea-level rise, will most likely make these stresses even more common and severe. To address these existing and potentially growing abiotic stresses in unfavorable environments, the combination of improved germplasm and adjusted crop and natural resource management (CNRM) options is necessary. Recent advances in germplasm development indicate considerable progress in submergence, salinity, and drought tolerance of new rice varieties. To complement these developments, better and more accompanying options for CNRM need to be developed and disseminated together with the new varieties.

Climate Change in Unfavorable Rice Environments

Climate change is an irregular and relatively slow process from a human perspective, even if it may be very fast from a geological point of view. Therefore, predictions of the effects of climate change are often given for the relatively distant future, say, for 2100 or beyond. But, for developing adjusted CNRM options, mainly changes in the near future are relevant, perhaps within the next 10-20 years. Also, likely developments in this shorter time frame can be more reliably estimated based on existing trends.

Increasing temperature is a potential threat to rice production because high temperature can affect rice production at all stages of development, particularly during fowering, when it causes spikelet sterility. It also increases plant respiration, affects photosynthesis, and shortens the grain-flling period, all of which contribute to lower grain yields. Very high temperatures (>33 °C) already occur in some Asian rice-growing regions, but they are mostly limited to the dry season when irrigated rice is grown. Critically high temperatures in the wet season are unusual, and high temperature alone is not very likely to become a major constraint in unfavorable rice areas within the next 20 years. However, even limited temperature stress can be aggravated by drought because the plant loses its ability to cool through transpiration, and the combination of both elements could cause more frequent temperature damage even in rainfed rice.

The same authors concluded that, in the near future, changes in precipitation may have a stronger effect on agricultural production than temperature changes. It is generally agreed that global climate change will cause a higher global average rainfall. Nevertheless, most studies assume that the effects of increased precipitation variability and intensity will rather increase the occurrence of drought. Already, it has been shown that the production of rice, maize, and wheat has declined in many parts of Asia in the past few decades because of increasing water stress, arising partly from increasing temperatures, increasing frequency of El Niño events, and a reduction in the number of rainy days.

In inland areas, more frequent high-intensity rains will cause more food events in the lower landscape portions, thereby increasing the occurrence of submergence events. And, coastal and deltaic regions will increasingly be affected by the already occurring sea-level rise. Although the likely average sea-level rise in the coming 20 years is not huge—mean estimates are between 32 and 62 mm for that period based on observed trends in the last century/decade—it is aggravated by an (often rapidly) sinking land surface in most big Asian deltas. Both trends together will affect the drainage of inland water; it might cause advancing salinity intrusion in rivers during the dry season and increase the threat from extreme weather events. Therefore, it is assumed that rice production in coastal and deltaic regions will increasingly be affected by submergence and salinity. Similarly, Allen et al concluded that the future challenge for irrigation systems in coastal/deltaic regions may primarily be the prevention of salinity intrusion and excessive fooding.

Drought-prone Environments

The biggest area of drought-prone rainfed lowland rice is located in India (13.3 million ha) and Thailand (8.2 million ha), but drought also regularly occurs in the other major rainfed lowland rice regions: Bangladesh, 5.1 million ha; Indonesia, 4.0 million ha; Vietnam, 2.9 million ha; Myanmar, 2.4 million ha; Cambodia, 1.6 million ha; and the Philippines, 1.3 million. Drought has long been recognized as the primary constraint to rainfed rice production. It can occur at any time during crop growth and is highly variable in space and in time. Drought can reduce rice yields directly by reducing transpiration or causing spikelet sterility or indirectly by impeding management operations such as crop establishment or weeding or by favoring low-input strategies.

One way to escape drought is to establish irrigation facilities. This approach was practiced in many previously purely rainfed lowland systems and contributed to large productivity increases. Examples are the establishment of large numbers of tubewells in, for example, eastern India and Bangladesh. Another possibility is the harvesting and storage of rainwater with examples in northeast Thailand, some regions of eastern India, the Philippines, and Indonesia, where considerable numbers of village and farm ponds for rainwater harvesting were.

If the drought pattern is relatively stable and of the early-season or late-season type, drought avoidance can be achieved by adjusting the cropping season to the time when water availability by rainfall is best or by reducing the length of the cropping season (and thereby the drought risk). This can be achieved with the help of shorter duration varieties or nonphotoperiod-sensitive varieties. Although these seem obvious options, many rainfed lowlands are still dominated by long-duration, photoperiod-sensitive variety types and no good shorter duration material is available to farmers.

Another important mechanism to moderate drought damage is improved nutrient management. Nutrient management is rarely seen as an option to mitigate drought stress, but nutrients are, in fact, often a limiting factor in rainfed lowlands. The resulting limited growth reduces access to available soil water resources and drought stress further reduces the plant availability of nutrients. Improved nutrition renders plants a stronger competitor for water and helps reduce unproductive water losses. These effects obviously cannot help in the case of extreme drought or drought around flowering, but medium drought events with considerable yield-reducing effects are common in many rainfed lowlands.

Another opportunity to reduce unproductive water losses and moderate drought effects is direct seeding. Dry direct seeding (DDS) allows earlier establishment than transplanting, thus reducing deep percolation and evaporation losses from early-season rains. The advantages were analyzed by Rathore et al for four different establishment methods of rainfed rice in eastern India. Sowing dry seed in dry soil could be undertaken before the rains started, while sowing dry seed in moist soil, as practiced with the biasi system, required 112 mm of water on average.

In comparison, crop establishment by transplanting (including, for that purpose, the necessary soil puddling) and wet plowing in the biasi system required approximately 500 mm of rainfall. Rice established by DDS on dry soil suffered less from water deficit, resulted in better rainfall use efficiency, and gave the best yields across the 5 years of the study. Research on minimum or zero-till options for rice is ongoing but no conclusive results are available yet.

Other suitable management methods tested in rainfed rice to increase the productive use of water and to reduce unproductive water losses are land leveling and soil improvement. Land leveling can be optimized with laser-guided equipment, but better field leveling can often be achieved with conventional farm equipment. Application and incorporation of organic or clayey materials can help increase soil water retention capacity, especially on coarse-textured soils. Where at least partial irrigation is possible, water-saving irrigation techniques such as alternate wetting and drying, saturated soil culture, and aerobic rice are available options to make best use of scarce water resources.

Submergence-prone environments Rice is the major crop in food-prone areas of South and Southeast Asia, providing food for millions of subsistence farming families. These areas are subject to either more frequent fash or temporary foods (submergence), longer term fooding of 20–50 cm (partial/ stagnant, semideep), deep water of >100 cm (deepwater rice), or very deep water of up to 3 or 4 m (foating rice). Rice productivity in these ecosystems is only about 1.5 t ha^{-1} because of the lack of high-yielding varieties tolerant of these stresses.

The challenges facing rice production in these food-prone areas are becoming more complex because of the continuous sea-level rise and the probable increase in extreme rainfall events caused by ongoing climate change, particularly in coastal areas where rice-based systems predominate. Efforts to improve rice productivity in these areas will ultimately contribute to global efforts for coping with these changes and ease their adverse effects on the food security of rice farmers and consumers.

Our current understanding of the constraints and challenges facing farmers in these areas is rapidly advancing, and prospects for increasing productivity are becoming increasingly visible. As an entry point, narrowing yield gaps in food-affected areas should begin with developing and deploying resilient, high-yielding varieties with submergence tolerance and shorter duration than the existing low-yielding, long-duration landraces.

These varieties will then provide great opportunities for proper crop management, more input use, and further adjustments to maximize system productivity and farmers' income. Good agronomic practices targeting proper crop establishment, better survival of fooded plants, and faster recovery thereafter already showed considerable promise in some areas. Soils in these areas are often fertile because of the recurring deposition of silt and organic materials carried in the foodwa-ter.

Freshwater resources are sometimes available from surface storage facilities and streams and, in some cases, from renewable underground water resources, providing opportunities for dry-season farming. This provides tremendous opportunities for food security in these areas, as both rice and other crops grown during the dry season are less vulnerable to climate perturbations commonly experienced in the wet season as a consequence of the cycling weather conditions or probable global warming, as witnessed in recent years with increasing incidences of cyclones and coastal storms.

In the past few decades, signifcant progress was made in developing high-yielding rice varieties adapted to rainfed and food-prone areas. However, research on developing good management strategies has been and is still lagging behind. For example, specifc nutrient recommendations have not been developed for food-prone areas and farmers often abstain from using inputs as a risk-aversion strategy. Also, traditional landraces normally beneft only a little from extra inputs, and available modern varieties were often heavily damaged when fooded, resulting in little or no gain from extra inputs. Increasing the availability of tolerant varieties provides more opportunities for developing and validating numerous management options effective in food-prone areas. Recommendations are being developed and refned and some of them are already being tested or outscaled in target areas.

Proper crop establishment is a major challenge in food-prone areas because rice is relatively more sensitive to fooding during germination and early seedling growth. Direct seeding is being practiced in many deepwater areas, and it has gained momentum in fash-food and other rainfed areas because of its lower cost and operational simplicity, besides the other ben-efts.

Our recent research identifed a few landraces that can germinate better in fooded soils, and proper seed and seedbed management practices were developed that can further enhance the performance of breeding lines obtained from these landraces. For transplanted rice, proper nursery management can signifcantly enhance crop establishment, survival, and recovery when fooding occurs shortly after transplanting. This can considerably increase grain yield if complete submergence occurs later during the vegetative stage. Such options should be particularly attractive to farmers since they need to apply them only on the small area occupied by seedbeds.

These options include proper nutrient management, use of organic manure, use of lower seed density, proper water management, and transplanting of older seedlings when fooding is anticipated early after transplanting. High nutrient application (especially N) should be avoided since it results in vigorous growth and less stored carbohydrate reserves needed for maintenance during submergence. In some food-prone areas, farmers practice double transplanting (even triple transplanting) to produce taller seedlings for transplanting in standing water at the beginning of the season (India and Bangladesh) or to rejuvenate seedlings while waiting for the foodwater to recede to levels that can allow transplanting in the main feld. Such a system with triple transplanting is practiced in Indonesia and could be further improved through the choice of proper varieties (e.g., photoperiod-sensitive varieties seem to be better adapted) or proper management of seedlings in nurseries or after transplanting in the feld.

Postsubmergence nutrient management, when possible, can also contribute to increased productivity. Application of nutrients that hasten recovery and increase early tillering can considerably increase yield because these early tillers can be productive. High early tillering also reduces the chances of excessive late tillering, which either considerably extends crop duration or leads to tillers that remain vegetative. Such post-fooding management is possible in areas where typical fash foods occur, after the water recedes to levels that make nutrient application possible. Responses to fertilizers, particularly N, can be observed, even if applied within a few days after the water recedes, and a second dose can be applied just before panicle initiation.

In some countries such as Vietnam, Bangladesh, India, and Thailand, dry-season rice became a major crop that transformed farmers' livelihood in food-prone systems, resulting in substantial improvements in rice production and food security. This has become possible after investing in irrigation and food control infrastructure, as in Vietnam, or through the use of subsurface water by shallow tubewells, as in Bangladesh. However, these resources are highly underexploited in some countries, as in Indonesia, in spite of the presence of renewable good-quality water at shallow depths. Dry-season rice could provide greater food security because it is less vulnerable to natural hazards, but the system requires initial investment in irrigation, development of suitable short-maturing varieties, and proper management options. This is also possible only if exercised at the community level to avoid the discouraging damage caused by rats and birds and high disease and insect pressure if just small areas are grown during the dry season. The availability of short-maturing, high-yielding varieties could also provide opportunities for various options of cropping patterns, particularly if combined with direct seeding of rice. However, the choice of nonrice crops in such systems should be carefully considered to ensure that they can ft within the available time window, that they have good market value, and that farmers have market access.

Salt-affected soils

Salinity and other associated stresses are important constraints to rice production in some soils in humid and subhumid coastal climates as well as in some inlands of Asia. Up to 27 million ha are believed to be affected to some extent by salt stress at the coasts of South and Southeast Asia, of which 3.1 million ha are in India, 2.8 million ha are in Bangladesh, and 2.1 million ha are in Vietnam.

The majority of these areas are not currently in use for agriculture, but some are potentially suitable for rice cultivation. Salinity in these areas varies seasonally, being high in the dry season due to capillary rise and peaking before the onset of rains. Salinity levels in soil and water then decrease progressively during the monsoon season and are lowest from June to September.

Unlike in the inlands, the dynamic nature of salinity in coastal areas makes it diffcult to conduct long-term soil reclamation. Rice is the most suitable crop for most of these coastal areas because of its ability to fourish in fooded soils, a condition necessary for leaching salts out of the soil.

One of the very likely consequences of global warming is the increase in area and severity of soil salinity, in both coastal and inland ecosystems. In coastal areas, an increase in salinity intrusion has already been witnessed in some of the low-lying deltas as in southern Bangladesh, Vietnam, and Myanmar. This is the consequence of catastrophic storm events, possibly aggravated by the slight increase in sea levels and the subsiding land surface in many deltas. In inland areas, salt deposition is possibly increasing as a consequence of increased evapotranspiration with rising temperatures.

Despite considerable progress made in developing rice varieties that are tolerant of salt stress, productivity of salt-affected areas still remains relatively low at 1–1.5 t ha^{-1}. However, yield in most of these areas can be raised by at least 2 t ha^{-1} when validated management and stress mitigation strategies are used together with salt-tolerant varieties. An array of technologies has been developed and found to be effective as amendments and mitigation measures for both short-term (coastal areas) and long-term (inland) reclamation of salt-affected soils.

These technologies include improved agronomic management packages, water harvesting and water management, and soil salinity and fertility management. The availability of short-maturing, salt-tolerant varieties also provides opportunities for crop intensifcation in areas where freshwater resources are available or could be made available through water harvesting and management of surface-water resources in the dry season. However, in areas where these resources are limited, considerable prospects may still exist through the introduction of short-maturing nonrice crops with less water requirements than dry-season rice.

Again, care should be taken when choosing these crops to cnsurc good market value and beneft to farmers as described above for food-prone areas. Developing the infrastructure that can restrict intrusion of saline water and enhance drainage such as food control polders and sluices can help ease the existing stresses and halt further degradation with global climate change. Improved management of natural resources should also take into account the indigenous knowledge of farmers in coping with these changes and their socioeconomic aspects, in parallel with the development of new concepts and strategies.

Because coastal areas of the tropics, particularly the low-lying deltas where rice is the dominant enterprise, are more vulnerable to any changes in sea-level rise or weather variability, diversifying farmers' options will minimize their vulnerability and risks. However, more information is needed on the extent of exposure to probable threats in order to formulate proper risk management strategies and policies for specifc agricultural ecosystems, including climate and food forecasting and warning. Exploring potentials to increase the value of farm output through intensifcation and increasing the production of higher value crops with less demand for water and better adaptation to existing stresses can also help uplift farmers' livelihoods, making them less vulnerable to any adverse consequences of seasonal or long-term climate perturbations.

Adjusting Cropping Systems to Cope with Climate Change

Greater variability in the onset of the monsoon, as a result of climate change, and increased incidence of either drought or high-rainfall events augur for greater uncertainty at the farm level. In response, there is a need for rice varieties with greater tolerance of abiotic stress (e.g., drought or submergence), but also for new options that give farmers greater fexibility to respond to the weather. The availability of technologies in the form of germplasm and CNRM options provides opportunities for intensifcation and also fexibility in the cropping calendar. Chea et al described "intensifcation" as growing more than one crop a year either by growing crops in different seasons or more than one crop in a season.

Options for rice-based cropping systems in lowlands that are available to farmers are, to a large extent, governed by the availability of adequate fresh water for crop growth and feld operations, such as soil tillage, and intervals when there is excess water and fooding. In these areas, options and decisions are governed by position in the to-posequence and resulting water fows and drainage capacities. "Windows of opportunity" in weather patterns allow farmers to prepare the land, establish the crop, and harvest the crop, depending on other system characteristics.

When only long-duration, photoperiod-sensitive varieties are available, and farmers have to puddle the land and transplant the crop to control weeds, opportunities to intensify are very limited. These are now greater, however, due to the availability of photoperiod-insensitive and short-duration varieties, improved market access for upland crops in some areas, and feasible direct-seeding options through mechanization and herbicides.

In Cambodia, for instance, a single rice crop is the most common cropping pattern in most lowland areas and varieties of early, medium, and late duration are grown on the upper, medium, and lower felds, respectively. Double cropping of rice in the lowlands is practiced with either early wet-season direct-sown rice followed by main-season transplanted rice or main-season rice followed by a dry-season rice crop.

Upland crops are grown either before or after rice. Intensifcation can improve food security and income, and Sengkea, as cited by Chea et al, showed that growing mungbean before rice where previously no prerice crop had been grown could increase farm income by 9%. In eastern India, Rathore et al reported that advancing the harvest of rice by dry direct seeding of rice rather than transplanting allowed for increased chance of a successful harvest of chickpea (*Cicer arietinum* L.) grown on residual moisture following rice. Similarly in northwestern Bangladesh, Mazid et al demonstrated that dry direct-seeded rice could give yields similar to those of

conventional transplanting and could advance the rice harvest by 7–10 days. Earlier harvest reduced the risk of terminal drought in rice and increased the chances of establishing a high-value crop such as chickpea or potato (*Solanum tuberosum* L.) on residual moisture. In this example, direct seeding of rice was made possible by either simple tools for line sowing and interrow weeding or herbicides. Similar examples of important changes in the cropping system as an adaptation to the environment and on the basis of new technologies were given for drought-prone, submergence-prone, and saline environments.

CNRM Options to Mitigate Climate Change

Rice felds contribute to global warming in several ways. They are estimated to contribute 10–15% of global methane emissions, and methane is a potent greenhouse gas. The reason is that, in fooded rice felds, organic matter (i.e., rice residues) is decomposed anaerobically, resulting in the formation of carbon dioxide and methane. Growing two or even three rice crops a year shortens the time of aerobic decomposition and increases the amount of crop residues, therefore further increasing methane emissions. In addition, intensive rice cropping generally uses considerable amounts of urea fertilizer, which is partly converted to nitrous oxide, another very potent greenhouse gas. Finally, the still widely practiced feld burning of residues contributes to the black carbon (soot) in the atmosphere, which was only recently identifed as a major contributor to global warming, especially in tropical Asia.

In traditional rice-livestock systems, still typical for many unfavorable rice production systems, a considerable part of the rice straw is removed as fodder for cattle and only a small part is returned to the felds as manure.

Straw is also used for various other purposes such as construction of houses and other shelters.

Usually, only one rice crop is grown, and only a little N fertilizer is used. Because of the low intensity and limited use of external inputs, straw (and grain) yields are relatively low in these systems. In addition, the felds are not fooded for most of the year, causing most organic matter to be decomposed aerobi-cally into CO_2 only. Therefore, the contribution of greenhouse gases from unfavorable/rainfed systems is limited.

However, the effect of intensifcation in these systems, as targeted by most of the technologies discussed above, is largely unknown and related research is needed. Also, little is known about the contribution of methane from cattle fed with rice residues.

In more modern, intensive systems, less straw is needed for animals because machines take their place. Simultaneously, intensifcation results in more straw being burned because it hinders soil preparation and crop establishment. Burning of rice residues in the feld has a range of effects on climate change. It produces considerable amounts of methane comparable with that produced during decomposition in anaerobic soils.

Second, as mentioned above, the soot released into the atmosphere contributes to global warming. But open-feld burning is also supposed to transform up to 3% of the initial biomass into black carbon or biochar, remaining on/in the soil. The transformation of organic matter into black carbon greatly reduces its degradability and could therefore contribute to a long-term carbon sink in rice soils. But, in recent times, residue burning was banned in most countries because it causes considerable air pollution and negatively affects public health, although it is still practiced in many places. In that situation, farmers are obliged to incorporate large amounts of straw into the soil, resulting in higher methane emissions.

Adjusted water, residue, and fertilizer management is therefore the main element that can be used to mitigate climate change. Mitigation options should probably be concentrated on intensive irrigated systems because most gains can be made. However, the target could also be to maintain low emissions of greenhouse gases from rainfed systems while simultaneously increasing their productivity. The main objective of methane-reducing management options is to maximize the aerobic decomposition of crop residues in and after the rice season, and/or to reduce the amount of rice residues remaining in the feld. This can be achieved, for example, by a change in water management, by soil preparation after the growing season instead of just before the next wet season, or by growing a nonrice crop in the dry season. A different approach would be to remove the residues completely and use them for energy generation, possibly even combined with carbon sequestration through biochar recycling. This approach would simultaneously reduce the emission of methane and soot resulting from feld burning.

To minimize nitrous oxide emissions, the soil needs to be kept anaerobic for about a week after N application. But N management should also avoid overfertilization and should be tailored to crop demand, depending on the site and the crop stage. Slow-release fertilizers could also play a role, provided that their value-cost ratio improves.

The change from rice-rice cropping systems to rice-nonrice systems could perhaps reduce greenhouse gas emissions further, but the effect on soil organic carbon could be counteref-fective because it has been shown that such a change can reduce soil organic matter concentrations signifcantly. A special case is the change to a cropping sequence that includes an N_2-fxing leguminous crop. Such crops can fx considerable amounts of N_2 that can contribute to increased nitrous oxide emissions.

These examples show that there are numerous interaction effects and that the consequences of management interventions within and beyond cropping systems are hardly understood. Therefore, considerable research is needed to determine the most promising mitigation options and to more fully comprehend the various effects of system management and CNRM on climate-relevant parameters.

Adaptation and Mitigation through Improved Technologies

The twin pillars under mitigation and adaptation are strategies

(i) mitigation and adaptation through novel technologies in crop production and management under projected climate change scenario and

(ii) sound governmental policy and political will to overcome the projected ill-effects of climate change in agriculture.

The first is by successful manipulation of the direct effects of climate change on grain crops, viz. reduction in duration, embryo abortion, spikelet sterility, effects on grain number and grain size, anthesis interval etc. The strategy involved here is the efficient use of conventional breeding and molecular/ mutation breeding by the use of biotechnological tools including marker assisted selection, whole genome expression analysis and its subsequent elucidation and gene finding by bioinformatics. The indirect effects, viz. decline in water resources, increased pests and disease incidence, loss of soil organic carbon should be tackled by conservation and efficient use of water, integrated pest management and conservation farming

Crop/cropping System-based Technologies

These will be mainly centered on promoting the cultivation of crops and varieties that fit into new cropping systems and seasons, development of varieties with changed duration that can over winter the transient effects of change, release of varieties for high temperature, drought and submergence tolerance, evolving varieties which respond positively in growth and yield to high CO_2.Besides varieties with high fertilizer and radiation use efficiency and also novel crops and varieties that can tolerate coastal salinity and salt water inundation are needed. Agricultural biodiversity and crop germplasm exploration for favorable traits is an important area that needs to be tapped to the fullest extent. Seeds, plants and plant parts exhibiting tolerance to temperature, water and other atmospheric stresses caused by climate change needs to collected and conserved to aid crop breeding research. A thorough revisit and re-evaluation of all the wild relatives, land races, extant varieties, modern varieties and breeding stocks could help in unraveling previously unknown or ignored traits than could prove more useful in the present scenario. Genetic resources could well prove to be the most important cost effective basic raw material which will allow agriculture to adapt to climate change.

Germination potential of transgenic seeds was several folds higher when challenged with salt and water stresses. In addition, they have remarkably robust root system in terms of root biomass and length. Strategies for genetic enhancement of heat tolerant genotypes especially in pulses by identifying and validating markers for high temperature tolerance coupled with yield potential is one of the key technological advances that can prove to be a significant strategy for adapting to climate change. An additional strategy is to take advantage of faster growth under higher temperatures, the new varieties, especially of the rabi cropping season should have characteristics of early flowering and early maturity and high produce.

Improved and novel agronomic and crop production practices like adjustment of planting dates to minimize the effect of high temperature increase induced spikelet sterility can be used to reduce yield instability, by avoiding flowering to coincide with the hottest period. Adaptation measures to reduce the negative effects of increased climatic variability as normally experienced in arid and semiarid tropics may include changing the cropping calendar to take advantage of the wet period and to avoid extreme weather events during the growing season. Crop varieties that are resistant to lodging may withstand strong winds during the sensitive stage of crop growth. In addition, improved crop management through crop rotations and inter cropping, integrated pest management, supplemented with agroforestry and afforestation schemes will be an important component in strategic adaptation to climate change in India. In grazing lands, pasture improvement is essential to combat impeding changes through planned grazing processes, enclosures for recovery, or enrichment planting.

Inter-cropping is an efficient strategy that can be followed with desirable outcome in the present climate change scenario. Grain legume intercrops have many potential benefits such as stable yields, better use of resources, weeds, pest and disease reductions, increased protein content of cereals, reduced leaching as compared to sole cropping systems. Establishment of seedbanks are of crucial importance in highly variable and unpredictable environments. This facility will provide a practical means to re-establish crops obliterated by major disasters and extreme climate events. This will also help in plant community dynamics, as differential plant germination strategies to buffer against inter-annual variability ingrowing conditions. The promotion of scientific agroforestry forms a key component in the war against climate change. Agroforestry systems buffer farmers against climate variability, and reduce atmospheric loads of greenhouse gases. Agroforestry can both sequester carbon and produce a range of economic, environmental, and socioeconomic benefits. For example, trees in agroforestry systems improve soil fertility through control of erosion, maintenance of soil organic matter and physical properties, increased Naccretion, extraction of nutrients from deep soil horizons, and promotion of more closed nutrient cycling.

Resource Conservation-based Technologies

The key resource conservation-based technologies are in situ moisture conservation, rainwater harvesting and recycling, efficient use of irrigation water, conservation agriculture, energy efficiency in crop production and irrigation and use of poor quality water. The suggested strategies are: characterization of biophysical and socioeconomic resources utilizing GIS and remote sensing; integrated watershed development; developing strategies for improving rainwater use efficiency through rainwater harvesting, storage, and reuse; contingency crop planning to minimize loss of production during drought/flood years. Zero tillage (ZT) has effectively reduced the demand for water in rice wheat cropping systems in more than 1 million ha of area in the Indo-Gangetic Plains. With ZT technology, farmers can realize higher yields and reduce production costs. In addition, ZT has a direct mitigation effect as it converts the green house gases like CO_2 into O_2 in the atmosphere and carbon, and enriches soil organic matter. Bed planting is widely adopted in the Indo-Gangetic Plains, proved to be a successful conservation technology. The main advantages are: increased water use efficiency, reduced water logging, better access for inter row cultivation, weed control and banding of fertilizers, better stand establishment, less crop lodging and reduced seed rates.

References

Adams, R.M., et.al. 1994. A reassessment of the economic effects of global climate change on U.S. agriculture. *Climatic Change* 30 (2): 147-167.

Gommes, R. 1993. Current climate and population constraints on world agriculture. In: *Agricultural Dimensions of Global Climate Change.*H.M. Kaiser and T.E. Drennen (eds.). pp. 67-86.

Kaiser, H.M. and Drennen T.E. (eds.). 1993. *Agricultural Dimensions of Global Climate Change.* St. Lucie Press, Delray Beach, Florida.

Pearman, G. 1988. *Greenhouse: Planning for Climate Change.* CSIRO, Canberra.

Rosenzweig, C., et.al. (eds.). 1995. *Climate Change and Agriculture: Analysis of Potential International Impacts.* ASA Special Publication No. 59. American Society of Agronomy, Madison, WI. 382 p.

Bibliography

Adams, R.M., et.al. 1994. A reassessment of the economic effects of global climate change on U.S. agriculture. *Climatic Change* 30 (2): 147-167.

Arnell, N.W. and Reynard, N.S. 1993. *Impacts of Climate Change on River Flow Regimes in the United Kingdom.* Institute of Hydrology, NERC, Water Directorate, DOE, July 1993.130 p.

Bazzaz, F., and W. Sombroek (eds.), *Global climate change and agricultural production. Direct and indirect effects of changing hydrological, pedological and plant physiological processes.* FAO and John Wiley & Sons, 345 pp.

Boden, T.A., Kaiser, D.P., Sepanski, R.J. and Stoss, F.W. (eds.). 1994. *Trends '93: A Compendium of Data on Global Change.* US Department of Energy, Carbon Dioxide Information Analysis Center, Oak Ridge National Laboratory, Oak Ridge, Tennessee.

Clark, W.G. & Hare, S.R. 2002. Effects of climate and stock size on recruitment and growth of Pacific halibut. *N. Amer. J. Fisher. Manag.,* 22: 852.

Clemmensen, C., Potrykus, A. & Schmidt, J. 2007. *Climate change and European fisheries.* European Parliament, policy department, structural and cohesion policies, fisheries. Brussels. Study.

Croxall, J.P., Trathan, P.N. & Murphy, E.J. 2002. Environmental change and Antarctic seabird populations. *Science,* 297: 1510.

Edwards, M. & Richardson, A.J. 2004. Impact of climate change on marine pelagic phenology and trophic mismatch. *Nature*, 430: 881.

Fajer, E.D. and Bazzaz, F.A. 1992. Is carbon dioxide a 'good' greenhouse gas? Effects of increasing carbon dioxide on ecological systems.*Global Environmental Change: Human and Policy Dimensions* 2: 301-310.

FAO. 1995. *World Agriculture: Towards 2010, An FAO Study.* N. Alexandratos (ed.). John Wiley, Chichester, UK, and FAO, Rome. 488 p.

Gommes, R. 1993. Current climate and population constraints on world agriculture. In: *Agricultural Dimensions of Global Climate Change.*H.M. Kaiser and T.E. Drennen (eds.). pp. 67-86.

Hadley Centre. 1992. *Transient Climate Change Experiment.* Hadley Centre, UK Met. Office, Aug. 1992. 20 p.

Hall, A.E. 1993. Breeding for heat tolerance. *Plant Breed Res.* 10: 129-168.

Houghton, J.T., Jenkins, G.J. and Ephramus, J.J. (eds.). 1990. *Climate Change: The IPCC Scientific Assessment.* Cambridge University Press, Cambridge.

Howarth, C.J. 1991. Molecular response of plants to an increased incidence of heat shock. *Plant Cell Environ.* 14: 831-841.

Idso, S.B., Kimball, B.A., Anderson, M.G. and Mouney, J.R. 1987. Effects of atmospheric CO2 enrichment on plant growth; the interactive role of air temperature. *Agric. Ecosystems & Environment* 20: 1-10.

IPCC. 1992. *Climate Change 1992.* The supplementary report to the IPCC scientific assessment. J.T. Houghton, B.A. Callandar and S.K. Varney (eds.). Cambridge University Press. Cambridge.

Jacobs, C.M.J. 1994. *Direct Impact of Atmospheric CO2 Enrichment on Regional Transpiration.* Thesis. Wageningcn Agricultural University. 179 p.

Kaiser, H.M. and Drennen T.E. (eds.). 1993. *Agricultural Dimensions of Global Climate Change.* St. Lucie Press, Delray Beach, Florida.

Kimball, B.A., Mauney, J.R., Nakayama, F.S. and Idso, S.B. 1993. Effects of increasing atmospheric CO2 on vegetation. *Vegetatio*104/105: 65-75.

Kukla, G. and Karl, T.R. 1993. Nighttime warming and the greenhouse effect. *Envir. Sci. Technol.* 27 (8): 1468-1474.

Kunst, L., Browse, J. and Somerville. 1989. Enhanced thermal tolerance in a mutant of *Arabidopsis* deficient in palmitic acid unsaturation.*Plant Physiol.* 91: 401-408.

Mahlman, J.D., 1997. Uncertainties in projections of human-caused climate warming. *Science,* 278:1416-1417.

McNaughton, K.G. and Jarvis, P.G. 1991. Effects of spatial scale on stomatal control of transpiration. *Agric. For. Meteorol.* 54: 279-301.

Morison, J.I.L. 1987. Intercellular CO2 concentration and stomatal response to CO2. In: *Stomatal Function.* E. Zeiger, G.D. Farquhar and I.R. Cowan (eds.). Stanford Univ. Press, California. pp. 229-251.

Parry, M.L., Carter, T.R. and Konijn, N.T. (eds.). 1988. *The Impact of Climate Variations on Agriculture. Vol. 1: Assessments in Cool Temperate and Cold Regions. Vol. 2: Assessments in Semi-arid Regions.* Kluwer, Dordrecht.

Pearman, G. 1988. *Greenhouse: Planning for Climate Change.* CSIRO, Canberra.

Rosenzweig, C. and Iglesias, A. 1994. *Implications of Climate Change for International Agriculture: Crop Modeling Study.* US Environmental Protection Agency. EPA 230-B-94-003, Washington DC.

Ruttan, W. (ed.). 1994. *Agriculture, Environment, Climate and Health: Sustainable Development in the 21st Century.* University of Minnesota Press, Minneapolis.

Seitz, F. 1994. *Global Warming and Ozone Hole Controversies, A Challenge to Scientific Judgement.* G.C. Marshall Institute, Washington DC. 32 p.

Smith, J.B. and Tirpak, D.A. (eds.). 1989. *The Potential Effects of Global Climate Change on the United States.* Report to Congress. EPA, Washington DC.

Tans, P.P., Fung, I.Y. and Takahashi, T. 1990. Observational constraints on the global atmospheric CO2 budget. *Science* 247: 1431-1438.

Thomas, R.B., Reid, C.D., Ybema, R. and Strain, B.R. 1993. Growth and maintenance components of leaf respiration of cotton grown in elevated carbon dioxide partial pressure. *Plant Cell Environ.* 16: 533-546.

Tyson, P.D. 1987. *Climatic Change and Variability in Southern Africa.* Oxford Univ. Press, Captetown. 220 p.

Vergara, B.S. 1980. Rice plant growth and development. In: *Rice Production and Utilization.* B.S. Luh (ed.). Avi Publishing, Westport, CT. pp. 75-86.

Wilson, C.A. and Mitchell, J.F.B. 1987. A doubled CO2 climate sensitivity experiment with a global model including a simple ocean. *J. Geophys. Res.* 92: 13315-13343.

Wittwer, S.H. 1995. *Food, Climate, and Carbon Dioxide: The Global Environment and World Food Production.* CRC Press/Lewis Publishers, Boca Raton, Florida.

Woodwell, G.M. and Mackenzie, F.T. (eds.). 1995. *Biotic Feedbacks in the Global Climatic System: Will the Warming Increase the Warming?* Oxford University Press, New York.

World Bank/UNDP/ADB/EC/French Government. 1993. *Sub-Saharan Hydro-logical Assessment.* July 1993. Washington DC. 17 p.

Yoshida, S. 1977. Rice. In: *Ecophysiology of Tropical Crops.* P.T. Alvim and T.T. Kozlowski (eds.). Academic Press, London. pp. 57-87.

Ziska, L.M. and Bunce, J.A. 1993. Inhibition of whole plant respiration by elevated CO2 as modified by growth temperature. *Physiol. Plant* 87: 459-466.